Manual of Small Public Water Supply Systems

U.S. EPA
Office of Drinking Water

CRC Press
Taylor & Francis Group
Boca Raton London New York

CRC Press is an imprint of the
Taylor & Francis Group, an **informa** business

First published 1992 by C. K. Smoley

Published 2019 by CRC Press
Taylor & Francis Group
6000 Broken Sound Parkway NW, Suite 300
Boca Raton, FL 33487-2742

First issued in paperback 2019

ISBN-13: 978-0-367-45029-8 (pbk)
ISBN-13: 978-0-87371-864-6 (hbk)

Visit the Taylor & Francis Web site at
http://www.taylorandfrancis.com

and the CRC Press Web site at
http://www.crcpress.com

Library of Congress Cataloging-in-Publication Data

Catalog record is available from the Library of Congress

PREFACE

Healthful, comfortable living requires the availability of an adequate supply of high quality water for drinking and domestic purposes.

Whenever feasible, a community should consider obtaining water from a public water system in order to enjoy the advantages of qualified supervision under the control of a responsible public agency. A public water system usually provides the best way to assure an uninterrupted supply of safe water.

This manual is primarily directed to the individual or institution faced with managing or operating a small public water system. This manual is an update to the EPA *Manual of Individual Water Supply Systems* which was published in 1982. EPA is issuing this revision of the manual to present current concepts and practices to the public water system owner. This is in response to the greater public awareness of health and environmental issues, and advances in water treatment practices.

EPA and others have noted a tremendous need for information and assistance to very small systems. These needs include engineering solutions, such as appropriate equipment to treat water and operations improvements. Non-engineering issues also have become a primary focus of concern which will require more attention in water supply improvement efforts. Among these concerns are proper financing and management of water supply improvements, community involvement in water supply, institutional support, and development of human resources for improved operations and management of water supplies.

The solutions offered in this manual are the result of the work of many U.S. agencies, individuals, and international organizations dedicated to the upkeep of safe and reliable water supply. In fact, many of the problems and solutions identified have been studied worldwide. Through this manual, EPA will be able to offer practical assistance to the many very small public water suppliers in the U.S., as well as those with similar operating conditions elsewhere, through such programs as the U.S. Peace Corps. This manual should be useful to other Federal agencies concerned with the development of small water supplies, and to State and local health departments, well drillers, contractors, as well as to owners and operators of any system which supplies drinking water.

As with the previous manual for individual and small water supplies, I hope that this manual will be a practical tool to many thousands of rural water systems in their enduring effort to supply safe and reliable drinking water.

ACKNOWLEDGMENT

This manual follows the general format of its predecessors: U.S. Public Health Service Publication No. 24, prepared by the Joint Committee on Rural Sanitation, and the U.S. EPA *Manual of Individual Water Supply Systems*, which was prepared under the direction of Mr. W. J. Whitsell.

Overall planning and management for the preparation of this manual was provided by Mr. Marc J. Parrotta, Environmental Engineer, U.S. EPA Office of Drinking Water. The support by those who prepared this manual which is technologically updated and more responsive to emerging environmental concerns is highly regarded.

Further, EPA acknowledges the valuable contributions of others who wrote and reviewed parts of this manual. EPA is especially indebted to the following who deserve particular acknowledgement for their contributions to this manual: Mark A. Thompson and Glenn M. Tillman of Malcolm Pirnie, Inc.; Frank Bell, Jr., Paul S. Berger, Walt Feige, Benjamin P. Smith, James E. Smith, Jr., and James J. Westrick of U.S. EPA; T. David Chinn of the American Water Works Association; Meg Harvey of ECOS, Inc.; and John Trax of the National Rural Water Association.

CONTENTS

Preface iii

Introduction
Overview of Small Public Water Systems 1
SDWA Requirements 1
General Requirements 2
Institutional Concerns 2
Operations 3

Part I - Selection and Management of a Water Source
Introduction 7
Rights to the Use of Water 7
Sources of Water Supply 7
Quality of Water 10
Safe Drinking Water Act (SDWA) 17
Water Testing and Labs 21
Contaminant Sources 21
Quantity of Water 22
Water Conservation 24
Fire Protection 25
Sanitary Survey 27

Part II - Ground Water
Rock Formations 33
Ground Water Basins 33
Sanitary Quality 34
Chemical and Physical Quality 34
Distance to Sources of Contamination 35
Evaluating Contamination Threats 36
Development of Ground Water 38
Construction of Wells 44
Sanitary Construction of Wells 58
Abandonment of Wells 63
Reconstruction of Existing Dug Wells 63
Special Considerations in Constructing Artesian Wells 64
Springs and Infiltration Galleries 64

Part III - Surface Water for Rural Use
Introduction 65
Sources of Surface Water 66
Development of Springs 74
Infiltration Galleries 76

Part IV - Water Treatment
Need and Purpose 79
Water Treatment 80
Disinfection 83
Disinfection with Ultraviolet Light 89
Disinfection with Ozone 90
Membrane Technologies 90
Aeration 91
Other Treatment 91
Package Plants 98
Household Water Treatment 100
Treatment Waste Disposal 103

Part V - Pumping, Distribution and Storage
Types of Well Pumps 105
Selection of Pumping Equipment 113
Sanitary Protection of Pumping Facilities 120
Installation of Pumping Equipment 121
Alternate Energy Sources and Pumps 123
Pumphousing and Appurtenances 126
Cross-Connections 136
Pipe and Fittings 138
Pipe Capacity and Head Loss 139
Protection of Distribution Systems 142
Disinfection of Distribution System 142
Determination of Storage Volume 143
Protection of Storage Facilities 144

Part VI - Information, Assistance and Community Support
Introduction 149
Information Resources 149
Community Involvement 150
Technical Assistance and Training 151
Financial Assistance 153
Water Rates 155
Personal Computers 157
Institutional Alternatives 157

Bibliography 161

Appendices:

A. Health Effects, Source of Contamination, and Treatment 167
B. Collection and Analysis of Bacteriological Samples 175
C. Identification by Human Senses 177
D. Recommended Procedure for Cement Grouting of Wells for Sanitary Protection 189
E. Emergency Disinfection 191
F. State Drinking Water Agencies 195
G. EPA Drinking Water Offices 199
H. National Organizations 201
I. Farmers Home Administration Offices 203

Index 207

LIST OF TABLES

Table *Page*

1. Maximum Contaminant Levels (MCLs) 18
2. Secondary Maximum Contaminant Levels (SMCLs) 19
3. Planning guide for water usc 23
4. Average household water use activities 24
5. Suitability of well construction methods to different geological conditions 43
6. Steel pipe and casing, standard and standard line pipe 53
7. Quantities of calcium hypochlorite and liquid calcium hypochlorite required for water well disinfection 61
8. Recommended mechanical analysis of slow sand filter media 82
9. Information on pumps 116,117
10. Allowance in equivalent length of pipe for friction loss in valves and threaded fittings 141
11. Seven-minute peak demand period usage 143
12. Tank selection chart-gallons 144

LIST OF ILLUSTRATIONS

Figure *Page*

1. The hydrologic cycle 8
2. Pumping effects on aquifers 40
3. Dug well with two-pipe jet pump installation 45
4. Different kinds of drive-well points 46
5. Well-point driving methods 47
6. Hand-bored well with driven-well point and "shallow well" jet pump 49
7. Drilled well with submersible pump 51
8. Well seal for jet pump installation 56
9. Well seal for submersible pump installation 57
10. Yield of impervious catchment area 67
11. Cistern 70
12. Pond 72
13. Schematic diagram of pond water-treatment system 73
14. Spring protection 75
15. Slow sand filtration diagram 81
16. Package Plant Diagram 99
17. Exploded view of submersible pump 107
18. "Over-the-well" jet pump installation 108
19. Typical solar pump systems 109
20. Typical wind powered pumps 110
21. Typical hand pumps 112
22. Typical air lift pump 113
23. Typical hydraulic ram 114
24. Determining recommended pump capacity 118
25. Components of total operating head in well pump installations 119
26. Vertical (line shaft) turbine pump mounted on well casing 122
27. Pumphouse 127
28. Clamp-on pitless adapter for submersible pump installation 129
29. Pitless unit with concentric external piping for jet pump installation 130
30. Weld-on pitless adapter with concentric external piping for "shallow well" pump installation 131
31. Pitless adapter with submersible pump installation for basement storage 132
32. Pitless adapter and unit testing equipment 135
33. Head loss versus pipe size 140
34. Typical concrete reservoir 146
35. Typical valve and box, manhole covers, and piping installations 147
36. Sources of Information and Assistance 152

Manual of Small Public Water Supply Systems

Introduction

OVERVIEW OF SMALL PUBLIC WATER SYSTEMS

Public demands for safe and dependable drinking water have a major effect on those who own or manage water supplies. Among the thousands of small public water systems in the U.S., many do not have the technical and information resources to meet the public's demands.

A *public water system* is one that serves at least 15 connections or at least 25 individuals. EPA regulations apply to public water systems, and should be followed because of the importance of water quality and the health risks associated with contaminants.

Water systems must provide high quality drinking water at a low cost. Essential to this task are: developing information sources, monitoring water quality, providing good system operation, and establishing preventive maintenance programs. Many owners of small water systems do not have the resources that they need to perform this task. However, some assistance is available from federal, state and municipal agencies as well as service organizations. Many of these organizations and the assistance they can provide are discussed in this manual as a practical source of information for the small public water system owner.

Information and public involvement are very important to successful water system operation. Technical and financial information from governmental and professional organizations can allow even the smallest water system to enjoy many of the benefits of larger systems, without the high cost. Starting an information-sharing network with the active support of water system managers and users can help provide an efficient, cost effective water supply system.

SDWA REQUIREMENTS

Passage of the federal Safe Drinking Water Act (SDWA) in 1974 allowed the federal government to establish national drinking water regulations, or rules, to protect public health.

Under the SDWA, the federal government develops national drinking water regulations to protect public health and welfare. Individual states are expected to carry out and enforce these regulations for public water systems. Public water systems must provide water treatment, as required, ensure drinking water quality through monitoring, and provide public notice of violations or possible contamination.

Congress expanded and strengthened the SDWA in 1986. The 1986 amendments require EPA to regulate more contaminants, define maximum contaminant levels, set compliance deadlines, regulate surface water treatment, remove lead, require disinfection and wellhead protection, and strengthen enforcement.

GENERAL REQUIREMENTS

Small public water system owners are responsible for the quality and availability of drinking water from the raw water source to the consumer. The public must be able to depend on the water system's management to provide a safe, reliable supply of drinking water at a reasonable cost. This responsibility requires owners to successfully manage all aspects of their water system and properly address the many problems faced by small water systems.

Another important requirement is training of water systems operators. Training may include proper maintenance procedures, chemical treatment and safety procedures. Such training is usually provided by states, consultants, or other organizations.

INSTITUTIONAL CONCERNS

Common problems faced by small water systems often involve difficulties obtaining information, funding, training and community support for operational needs and capital improvements.

Information Sources

State, county or local health departments are the best contacts for information on water quality, regulations, or other specific concerns. Other sources are also available, such as the EPA Safe Drinking Water Hotline (1-800-426-4791) to answer questions about requirements under the SDWA. National organizations such as the American Water Works Association, National Rural Water Association and National Water Supply Improvement Association can also provide technical assistance with specific problems, and contacts with water equipment vendors.

Pamphlets, journals, seminar notes and other written information on various topics are available from the organizations listed above. Contacts within these organizations may help solve particular problems. Information-sharing networks are an important part of transferring technical knowledge within the water supply industry. Small public water system owners can benefit from organizing an active information-sharing network. Information sources are discussed in greater detail in Part VI of this manual.

Financial Options

Many small water systems often cannot afford the interest costs of standard loans, or cannot afford other upgrades and changes that some loan or grant programs require. This is especially true for small water systems that have only a few customers. Loans and grants intended especially for these systems are available from several sources including: the Farmers Home Administration; U.S. Department of Housing and Urban Development; U.S. Small Business Administration; and several other government and private programs. These are discussed further in Part VI of this manual.

Public Relations/Community Support

A public relations program conducted by a water system not only educates and involves the public, but encourages trust and support. People are interested in things that affect them, and clean drinking water plays an important role in providing a high quality of life. A community needs to know about drinking water issues, and a water system owner may help increase the public's awareness and understanding. This encourages a broad-based, informed and supportive group within the community that may assist in obtaining the improvements needed to provide a safe, reliable water supply.

OPERATIONS

Treatment

Without treatment, water from natural sources may not be suitable for drinking. The type of water treatment needed depends on the chemical, physical and biological makeup of the water. When studying a potential treatment system, operation and maintenance costs must be considered, as well as the capital (or initial equipment) cost. Operation and maintenance costs may vary greatly between different water systems providing the same type of treatment. This may have an impact on the cost to the water user.

Preventive Maintenance

One of the most neglected aspects of water supply systems is proper maintenance. Crisis maintenance, or correcting system failures when they are discovered, is a common practice. However, this creates higher costs in the long term and less protection for the drinking water supply. Proper maintenance includes a good preventive maintenance program that can help prevent costly system failures. This is always true, from the single home owner to the largest water supply system.

Case Studies

The following case studies are provided to show how communication and cooperation between owners of small public water systems and regulatory agencies, professional organizations, and water customers can solve some basic water concerns.

Case Study 1

Problem: The concentration of fluoride in a small public water system exceeds the maximum contaminant level of 4.0 mg/L.

Discussion: The system owner contacts the state agency in charge of water quality about possible treatment methods and cost of fluoride reduction. The treatment choices discussed include reverse osmosis, electrodialysis reversal and activated alumina. In addition, they looked for other water supply sources with lower levels of fluoride.

During their research, the owner discovered that the county is planning to extend its water distribution system to within 1,200 feet of their community. Comparing the cost of fluoride removal treatment to the cost of connecting to the neighboring water system shows that connection to the county supply would be less costly.

Solution: The water supply containing high levels of fluoride is abandoned, and the users are connected to the new extension of the county supply.

Case Study 2

Problem: A water system that supplies a trailer park finds that its shallow well system usually delivers cloudy water after a heavy rainfall.

Discussion: The State Health Department studies the problem and finds that the water supply is influenced by surface water. Surface water may transmit harmful microorganisms to the people who drink the water. Therefore, the water should be filtered. Homeowners, local health officers and the trailer park manager meet to discuss these findings and decide how to treat the water. A small package water treatment plant using sand filters is discussed as a low-cost answer.

Solution: Vendors for supplying a small package plant are found and cost is discussed. The package filtration plant is installed and regular visits by a contract operations and maintenance engineer are set up to keep the treatment system working properly.

Case Study 3

Problem: A small public water system regularly exceeds the standards for coliform bacteria in water.

Discussion: System owners have been unable to solve the problem, and agree that outside help is needed. The water system manager asks the county health department and a local professional engineer for help. After a review of various alternatives to provide disinfection, several reliable and low-cost methods of simple chlorination are suggested.

Solution: The system buys and installs a simple solution-feed chlorinator.

Case Study 4

Problem: Several households connected to a public water system have been tested and found to contain high lead levels at their tap.

Discussion: After a careful inspection, it is found that the source of lead is the lead solder used in their household plumbing.

Solution: To increase awareness and understanding of this problem, the water system owner sends a notice providing information on lead in water and a copy of an EPA pamphlet to all of the customers. Homeowners are also advised that flushing their water lines every morning or after extended periods of non-use will keep lead levels in the water as low as possible.

The system owner explores several ways to reduce the water's tendency to dissolve the lead in pipes and solder, including chemical corrosion inhibitors.

A system is installed at the treatment plant to add a chemical that controls the water's corrosivity and lessens leaching of lead from the household plumbing.

Part I

Selection and Management of a Water Source

INTRODUCTION

Planning a water supply system requires that the quality of the water and available sources be evaluated. In addition, a basic knowledge of water rights and the hydrological, geological, chemical, biological, and possible radiological factors affecting the water is helpful. These factors are usually interrelated because water or water vapor continually circulates from the oceans to the air, over the surface of the land and underground, and back to the oceans. This circulation is called the hydrologic cycle (see Figure 1).

RIGHTS TO THE USE OF WATER

The rights of an individual to use water for household, irrigation, or other purposes varies in different states. Some water rights come from owning the land bordering on or overlying the source, while others are acquired by meeting certain legal requirements. There are three basic types of water rights.

Riparian

These are rights acquired with the title to the land bordering on or overlying the source of water. Whether or not a riparian right exists depends on the laws of each state.

Appropriative

These are rights acquired for the use of water by following a specific legal procedure.

Prescriptive

These are rights acquired by diverting and putting to use, for a period specified by statute, water to which other parties may or may not have prior claims. The procedure necessary to obtain prescriptive rights must conform with the water-rights laws of each individual state.

A property owner should consult the appropriate state legal authority and clearly establish his rights to the use of water.

SOURCES OF WATER SUPPLY

At some time in its history, the oceans contained all water. By evaporation, moisture is transferred from the ocean surface to the atmosphere, where winds carry the moisture-laden air over land. Under certain conditions, this water vapor condenses to form clouds, which release their moisture in the form of rain, hail, sleet, or snow.

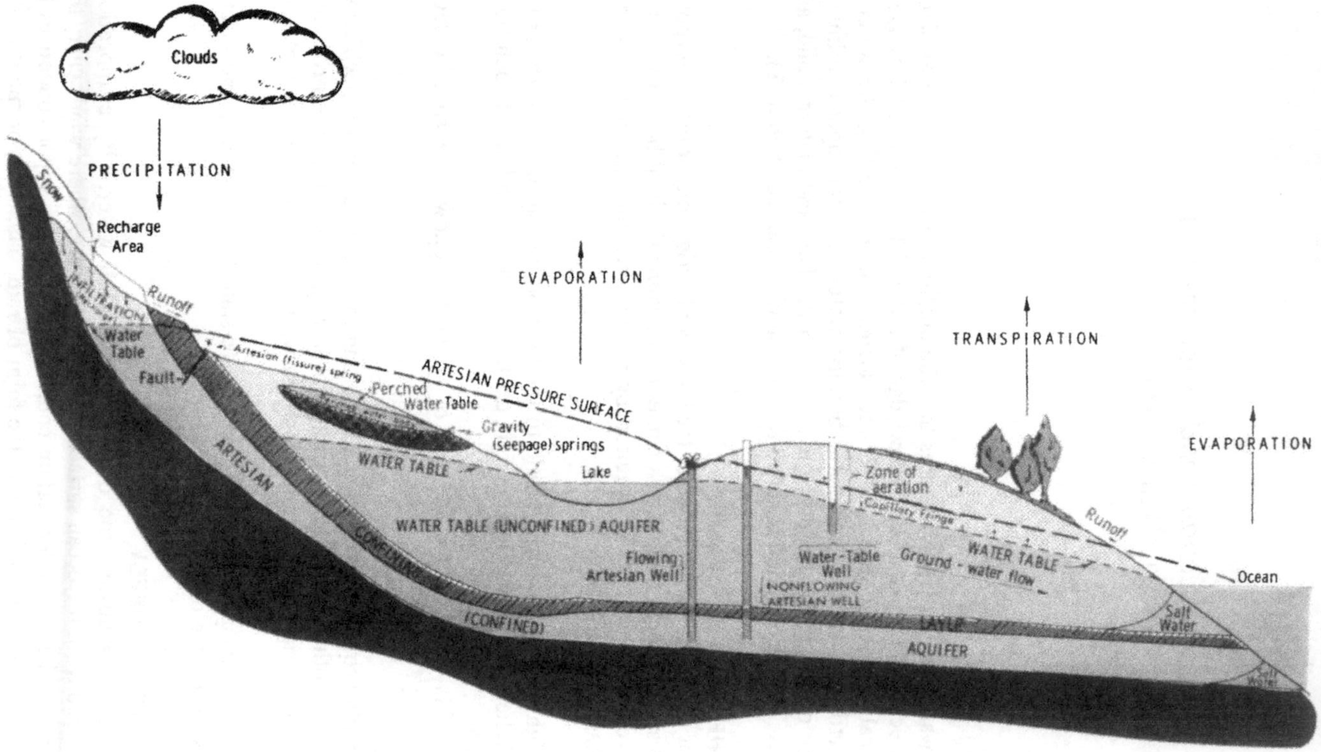

FIGURE 1. - *The hydrologic cycle.*

When rain falls, part of it may evaporate and return immediately to the atmosphere. Moisture in excess of the amount that wets a surface or evaporates is available as a potential source of water supply.

Ground Water

Some of the precipitation may seep into the soil (see Figure 1). This water replenishes the soil moisture or is used by growing plants and returned to the atmosphere by transpiration (water vapor released to the air by plants). Water that moves downward (percolates) below the root zone reaches a level at which all of the openings or voids in the ground are filled with water. This zone is known as the "saturation zone." Water in the saturation zone is referred to as "ground water." The upper surface of the saturation zone, if not restricted by an impermeable (watertight) layer, is called the "water table." When the ground formation over the saturation zone keeps the ground water at a pressure greater than atmospheric pressure, the ground water is under "artesian pressure."

The name "artesian" comes from the ancient province of Artesium in France, where, in the days of the Romans, water flowed to the surface of the ground from a well. Not all water from artesian wells flows above the surface of the land. An artesian well is one in which the water rises above the top of the aquifer. An aquifer, or ground water reservoir, is an underground layer of permeable rock or soil that permits the passage of water.

The porous material just above the water table may contain "capillary" water in the smaller void spaces. This zone is referred to as the "capillary fringe." It is not a true source of supply, however, since the water held here will not drain freely.

Because of the irregularities in underground formations and in surface topography (hills and valleys), the water table occasionally meets the surface of the ground. As a result, ground water moves to these locations and seeps out of the aquifer into a stream, spring, lake or ocean. Ground water is continually moving within the aquifer, even though the movement may be very slow. The direction of the ground water flow, especially when affected by well pumping, does not necessarily follow the surface slopes. The water table, or artesian pressure surface, slopes downward from areas of recharge (water inflow) to lower areas of discharge (water outflow). The differences in these slopes causes the ground water to flow within the aquifer. Seasonal variations in the supply of water to the aquifer, such as snowmelt, can cause considerable changes in the elevation and slope of the water table and the artesian pressure level.

Wells

A well that penetrates the water table can be used to extract water from an aquifer. As water is pumped from the well, the water table near the well is lowered. If pumping continues at a faster rate than the recharge of the water table, the "sustained yield" of the well is exceeded. If the sustained yield is exceeded for a long period of time, the aquifer may become depleted or other undesired results may occur. In that case, the "safe yield" of the aquifer has been exceeded. For example, salt-water may enter the aquifer where wells are near the seashore or other saline waters.

Springs

An opening in the ground from which ground water flows is a spring. Water may flow by force of gravity (from water-table aquifers), or be forced out by artesian pressure. The flow from a spring may vary considerably with changes in the water-table or artesian pressure. For further discussion, see Part II.

Surface Water

Surface water collects mainly as a result of direct runoff from precipitation. Precipitation that does not seep into the ground or evaporate, flows over the ground surface and is called direct runoff. Direct runoff flows over surfaces such as stream channels and other natural or artificial runoff channels.

In some areas, a source of water is the rainfall collected on the roof surfaces of homes, barns, and other buildings. Water from those surfaces can be collected and stored in tanks called cisterns. In some instances, natural ground surfaces can be conditioned to make them impermeable. This conditioning will increase runoff to cisterns or large artificial storage reservoirs, thereby reducing loss by seepage into the ground.

Runoff from the ground may be collected in either natural or artificial reservoirs. A portion of the water stored in surface reservoirs evaporates and seeps into the ground water table through the pond bottom. Transpiration from vegetation in and adjacent to ponds constitutes another means of water loss.

Ground and Surface Water

Ground water may become surface water at springs or at intersections of a water body and a water table. During extended dry periods, stream flows consist largely of water from the ground water reservoir. As the ground water reservoir is drained by the stream, the flow will reach a minimum or may cease altogether. It is important in evaluating stream and spring supplies to consider seasonal changes in flow that may occur due to snow or dry periods.

Snow

Much of the snow falling on a watershed remains on the ground surface until temperatures rise above freezing and it melts. In the mountainous areas of the western United States, snow storage is an important source of water supply through much of the irrigation season. Water supply systems in these areas are usually helped by actions which increase the snowpack and reduce melting.

Alternate Sources

Another possible source of water supply is an existing water system which may service a nearby community. This may be an attractive, low-cost solution, especially considering the costs of developing and operating a new water supply.

QUALITY OF WATER

Precipitation in the form of rain, snow, hail, or sleet contains very few impurities. It may contain tiny amounts of minerals, gases, and other substances as it forms and falls through the earth's atmosphere. However, it has virtually no bacterial content.

Once precipitation reaches the earth's surface, however, there are many chances for mineral and organic substances, micro-organisms, and other forms of pollution (contamination)[1] to enter the water. When water runs over or through the ground surface, it may pick up particles of soil. This is noticeable in the water as cloudiness, or "turbidity". It also picks up particles of organic matter and bacteria. As surface water

---

1 **Pollution as used in this manual means the presence in water of any foreign substances (organic, inorganic, radiological, or biological) which tend to lower its quality to a point that it constitutes a health hazard or impairs the usefulness of the water. Contamination, where used in this manual, has essentially the same meaning.**

seeps through the ground to the water table, most of the suspended particles (even some bacteria) are filtered out. However, the water may pick up additional contaminants or particles when it comes in contact with underground mineral deposits.

The widespread use of man-made chemicals, including pesticides, herbicides, insecticides, and industrial and medical chemicals, has caused a renewed interest in the quality of water. Many of these materials are known to be toxic and/or carcinogenic (cancer-causing) to humans and may have other undesirable characteristics. Water pollution has also been traced to sewage or waste water sources containing synthetic (man-made) detergents. Chemical and bacteriological analyses to determine water quality may be performed by a state or local health department or by a commercial laboratory.

Characteristics that describe water as it moves over or below the ground may be classified under four major headings: physical, chemical, biological and radiological. Analytical methods for these parameters are discussed below and can be found in *Standard Methods for the Examination of Water and Wastewater*[2].

Physical Characteristics

Physical characteristics relate to the appearance of water, its color or turbidity, taste and odor, temperature and foamability. The water, as used, should look, taste and smell clean.

Color. Dissolved organic material from decaying vegetation and certain inorganic matter causes water to appear colored. Occasionally, algae blooms or the growth of aquatic micro-organisms may also give color. Iron and manganese can cause a red or black color in water and may stain bathroom fixtures. While color itself does not usually mean that the water is not safe, its presence is unpleasant and suggests that the water needs treatment. Typically, treated water will have a measurement of less than five "color units." Color may also indicate mine drainage or the presence of decaying organic matter, which may form harmful trihalomethanes during disinfection with chlorine.

Foamability. Foam in water is usually caused by concentrations of detergents greater than 1 mg/L. The user should understand that if enough detergent is in the water supply to cause a noticeable froth to appear on a glass of water, hazardous materials from sewage or other pollution may also be present. Foaming is also undesirable in drinking water because it looks unclean.

Taste and Odor. Taste and odor in water can be caused by foreign matter such as organic compounds, inorganic salts, or other dissolved minerals or gases. These materials may come from household, industrial, agricultural, or natural sources. Drinking water should not have any unpleasant taste or odor.

Tastes and odors in water are usually tested for both strength and type. A sample is collected in glass containers and tested as soon as possible. The test can be delayed for a few hours if refrigeration is available.

A simple test consisting of taking a sample and observing it for clarity and smelling it for possible odors can be an early sign of pollution. If there is a detectible smell or taste, the "threshold odor test" can then be performed at a laboratory to determine the

---

2 American Public Health Association, American Water Works Association, and Water Pollution Control Federation, *Standard Methods for the Examination of Water and Waste Water*, 17th edition, Amer. Pub. Hlth. Assn., Washington, D.C. (1989)

severity of the problem. If pollution is suspected, other tests should be done and advice from local county and state agencies obtained.

Temperature. The most desirable drinking waters are consistently cool and do not have temperature changes of more than a few degrees. Ground water and surface water from mountainous areas generally meet this standard. Most individuals find that water having a temperature between 50°F (10°C) and 60°F (16°C) is most palatable. Temperature may also increase sensory response to tastes and odors.

Turbidity. The presence of suspended material such as clay, silt, finely divided organic material, plankton, and other inorganic material in water is known as turbidity. Turbidity in excess of five units can be easily seen in a glass of water, and is usually undesirable because it causes the water to look unclean. Treated water normally has turbidity levels of less than 0.5 NTU[3].

Clay or other inert suspended particles in drinking water may not adversely affect health, but water containing such particles may require treatment to make it suitable for use and to increase disinfection effectiveness. Turbidity's major danger in drinking water is that it can harbor bacteria, require the use of more chlorine during disinfection, or react with chlorine to form harmful by-products. Surface water supplies are generally more turbid than ground water, which is normally less than 1 NTU. Variations in the ground water turbidity following a rainfall may be considered a sign of surface water infiltration or other pollution since ground water turbidity does not usually change.

Chemical Characteristics

The materials that form the earth's crust affect not only the quantity of water that may be taken from a well, but also its chemical characteristics. As surface water seeps downward to the water table, it dissolves portions of the minerals contained in soils and rocks. Ground water, therefore, usually contains more dissolved minerals than surface water.

The chemical characteristics of water in a particular place can sometimes be predicted from analyses of nearby water sources, unless a local source of pollution is present. Chemical and other water quality data are often available in published reports of the U.S. Geological Survey or from Federal, State, and local health, geological, and water agencies. If information is not available, a chemical analysis of the water source should be made. Some state health and geological departments, as well as state colleges, and many commercial laboratories may provide this service. Portable water test kits are also available. Knowing about the chemical quality of a water supply source is important in order to determine what treatment, if any, is required.

Some substances, when present in water, may corrode, or wear away parts of the water system or stain fixtures and clothing. Chemical analysis can detect those substances. Proper sample size and the sample collection methods are important; the testing facilities instructions should be followed.

The following is a discussion of some of the chemical characteristics of water.

Alkalinity. Alkalinity results from bicarbonate, carbonate, or hydroxide compounds in the water. The presence of alkalinity is determined by standard methods involving titration with various indicator solutions. Knowledge of the compounds causing alkalinity

---

[3] NTU - Nephelometric Turbidity Unit, a standard unit of measurement for turbidity

is useful in selecting chemical clarification, softening, and corrosion control procedures for water supplies.

Aluminum. Aluminum may, even at low levels (0.05 mg/L), cause precipitation and increased turbidity in water as it passes through a distribution system. The World Health Organization recommends that levels of aluminum in water not exceed 0.2 mg/L because of the increased potential for water discoloration. EPA has recommended that States set aesthetic standards of 0.05 to 0.2 mg/L to prevent aluminum precipitation or discoloration of drinking water. The level chosen by a State is determined by water quality and treatment considerations.

Chloride. Most waters contain some chloride in solution. Sources include the leaching of marine sediments, intrusion of sea water, and industrial and domestic wastes. Chloride concentrations in excess of 250 mg/L usually produce a noticeable taste in drinking water. In areas where the chloride content is higher than 250 mg/L and all other criteria are met, however, it may be necessary to use a water source that exceeds this limit.

A sudden increase in the chloride content of a water may indicate pollution from sewage sources, particularly if the normal chloride content is known to be low.

Copper. Copper is found in some natural waters, particularly in areas where copper ore deposits have been mined.

Corrosive water passing through copper pipes may also pick up excessive amounts of copper. Copper in small amounts is not considered unhealthy, but produces an unpleasant taste. A larger toxic dose of copper causes nausea, and prolonged ingestion may result in liver damage. For this reason, the recommended limit for copper is 1.0 mg/L.

Hardness. Hard water and soft water are relative terms. Hard water retards the cleaning action of soaps and detergents, causing additional expense in the form of extra work and cleaning agents. Furthermore, when hard water is heated it will deposit a hard scale (as in a kettle, heating coils, or cooking utensils) which damages hot water pipes and requires additional heating fuel or energy expenditures.

Calcium and magnesium salts, which are the major cause of hardness in water supplies, are divided into two general classifications: carbonate, or temporary hardness, and noncarbonate, or permanent hardness.

Carbonate, or temporary hardness is so called because heating the water results in its removal. When water is heated, bicarbonates break down into solid particles that stick to a heated surface and the inside of pipes.

Noncarbonated, or permanent, hardness is so called because it is not removed when water is heated. Noncarbonated hardness is due largely to the presence of the sulfates and chlorides of calcium and magnesium in the water. Water with total hardness concentrations from 0 to 75 mg/L is considered soft, 75 to 150 mg/L is considered moderately hard, 150 to 300 mg/L is considered hard, and over 300 mg/L is considered very hard.

Fluorides. In some areas, water sources contain natural fluorides. Where the concentrations approach optimum levels (0.7 to 1.2 mg/L) health benefits have been observed. In such areas, the incidence of dental cavities has been found to be below the rate in areas without natural fluorides.[4]

Excessive fluoride (greater than 4 mg/L) in drinking water supplies may cause fluorosis (mottling) or discoloration of teeth, which increases as the fluoride level rises. When fluoride concentrations exceed 4.0 mg/L, the risk of skeletal fluorosis or other adverse health effects is increased.

Iron. Small amounts of iron are often found in water because of the large amount of iron present in the soil, and because corrosive water will pick up iron from pipes. Clothing washed in water containing excessive iron may become stained a brownish color. Iron also affects the taste of beverages such as tea and coffee. The recommended limit for iron is 0.3 mg/L.

Lead. Lead accumulations in the body and prolonged exposures to even very small amounts, particularly among young children and pregnant women, can result in serious health effects. The effects include delays in normal physical and mental development and impairment in learning and behavior. In 1991, EPA issued a regulation for public water supplies which includes an "action level" of 0.015 mg/L for lead in drinking water measured at household water taps. A public supply would have to take several steps to minimize lead in drinking water if 10 percent or more of the homes served by the supply exceed the lead "action level".

Excessive lead is occasionally found in source water, but the usual cause of excessive lead in tap water is lead in household plumbing. Solder and fixtures containing lead can leach lead into drinking water. In 1986, the U.S. Congress banned the use of lead solder containing greater than 0.2 percent lead and restricted the lead content of faucets, pipes and other plumbing materials to 8.0 percent. Nevertheless, materials containing lead are present in water distribution systems and in millions of homes.

Manganese. Manganese produces a brownish color in laundered clothing, produces black particles on fixtures, and affects the taste of beverages, including coffee and tea. The recommended limit for manganese is 0.05 mg/L.

Nitrates. Nitrate can cause the blood disorder methemoglobinemia (infant cyanosis or "blue baby disease") in infants who have been given water or fed formulas prepared with water having high nitrates. A household water supply should not contain nitrate concentrations in excess of 45 mg/L (10 mg/L expressed as nitrogen). Excess nitrate levels in water may be caused by contamination from livestock manure or from nitrogen fertilizer applied to farmland. Wells may also be contaminated by lawn fertilizers used by homeowners.

In some polluted wells, nitrite is also present in water and is even more hazardous, especially to infants. When a high nitrite concentration (i.e., at levels greater than 1 mg/L) is found, the water must not be used for feeding infants. The nitrite concentration should be determined, and if too high, advice obtained from health authorities about the safety of the water for drinking.

---

4 The addition of about 1 mg/L of fluoride to water supplies has been found to help prevent tooth decay in children. Some natural water supplies already contain amounts of fluoride that exceed the recommended optimum concentrations.

Pesticides. Careless use or improper storage of pesticides can contaminate water sources and make the water unsafe for drinking. Numerous cases have been reported of well contamination resulting from termite-control treatment. Pesticides must not be used near wells.

Other synthetic (man-made) organic chemicals, some of which are used by industries for many purposes, are also of concern in the protection of drinking water. These organic chemicals can be hazardous to human health or cause taste and odor problems in drinking water.

pH. pH is a measure of the hydrogen ion concentration in water. It is also a measure of the acid or alkaline content. The pH scale ranges from 0 to 14. Seven represents neutrality, while values less than 7 indicate increasing acidity and values greater than 7 indicate increasing alkalinity. The pH of water in its natural state is generally in the range of pH 5.5 to 9.0. Determination of the pH value assists in the control of corrosion, the determination of proper chemical dosages, and adequate control of disinfection.

Silver. Silver in water at levels in excess of 0.1 mg/L can cause a permanent gray discoloration of the skin, eyes, and mucous membranes known as argyria or argyrosis.

Sodium. When it is necessary to know the precise amount of sodium present in a water supply, a laboratory analysis should be made. Ion-exchange water softeners will increase the amount of sodium in drinking water. For this reason, water that has been softened should be analyzed for sodium whenever a person needs to keep a precise record of their sodium intake. For a healthy person, the sodium content of water is not a concern because the intake from table salt in the ordinary diet is so much greater, but for those on a low-sodium diet, sodium in water must be considered. The usual low-sodium diets allow for 20 mg/L sodium in the drinking water. When this limit is exceeded, a physician's advice should be sought.

Sulfates. Waters containing high concentrations of sulfate (greater than 250 mg/L) caused by the leaching of natural deposits of magnesium sulfate (Epsom salts) or sodium sulfate (Glauber's salt) may be undesirable because of their laxative effects.

Total Dissolved Solids (TDS). TDS refers to the presence of ions such as sodium, chloride, sulfate, and calcium in water. Some of these dissolved solids may cause taste and odor problems, deteriorate plumbing and appliances, or cause mineral precipitation. Because water with high TDS levels tastes bad, a maximum level of 500 mg/L is recommended in drinking water supplies.

Zinc. Zinc is found in some natural waters, particularly in areas where zinc ore deposits have been mined. Zinc is not considered detrimental to health, but it will impart an unpleasant taste to drinking water. For this reason, the recommended limit for zinc is 5.0 mg/L.

Serious surface and ground water pollution problems have developed from existing and abandoned mining operations. Among the worst are those associated with coal mine operations, where heavy concentrations of iron, manganese, sulfates, and acids have resulted from the weathering and leaching of minerals.

Biological Characteristics

The presence of biological organisms in water are very important to public health considerations and can significantly change the physical and chemical characteristics of water. Water for drinking and cooking must be made free from disease-producing organisms.

Total Coliform as Pollution Indicator. Organisms which are known to cause waterborne disease include bacteria, protozoa and viruses. Some algae and helminths (worms) may also be capable of producing disease. Symptoms of waterborne disease may include diarrhea, cramps, nausea and possibly jaundice.

Unfortunately, specific disease-producing organisms present in water are not easily identified. It would be very difficult, expensive and time-consuming to monitor for each of these organisms. For this reason, it is necessary to select an easily measured "indicator organism," whose presence indicates that disease-producing organisms may be present. A group of closely related bacteria, the total coliform, have been selected as an indicator of harmful organisms in drinking water.

Total coliforms are common in the water environment. One of them, *Escherichia coli (E. coli)* often referred to as fecal coliform) only grows in the intestine of warm blooded animals. Total coliform bacteria are generally not harmful. Water treatment should remove any total coliforms from water entering the distribution system. The presence of these bacteria in drinking water generally is a result of a problem with water treatment or the pipes which distribute the water, and indicates that the water may be contaminated with harmful organisms. Total coliforms are not only a useful indicator of potential sewage contamination, but are also a useful screen for the actual presence of *E. coli*. The presence of *E. coli* is strong evidence that fresh sewage is present.

Total coliform are not a perfect indicator of the actual or potential presence of harmful organisms, because some disease-producing organisms, especially protozoa such as *Giardia* and *cryptosporidium*, are able to withstand treatment which remove total coliforms. These two protoza are often found in surface waters which are contaminated by human sewage or wildlife, which are principle carriers of these organisms.

Other Biological Factors. Certain forms of aquatic plants and microscopic animal life in natural water may be either stimulated or retarded in their growth by physical, chemical, or biological factors. For example, the growth of algae, minute green plants usually found floating in surface water, is stimulated by light, heat, nutrients such as nitrogen and phosphorus, and the presence of carbon dioxide as a product of organic decomposition. Their growth may, in turn, be retarded by changes in pH (measure of acidity), the presence of inorganic impurities, excessive cloudiness or darkness, low temperature, and the presence of certain bacteria.

Continuous cycles of growth and decay of algae in water may result in the production of undesirable by-products that may adversely affect water quality. The same is true of certain non-harmful bacteria or other microorganisms that live in natural waters.

A water source should be as free of biological activity as possible. Select water sources that do not normally support much plant or animal life, and protect these sources from contamination by biological and fertilizing agents. Stored water should not be exposed to excessive light or temperature. Disinfection may be necessary to properly control biological activity.

Radiological Characteristics

Radiological factors must be considered in areas where there is a possibility that the water may have come in contact with radioactive substances.

Radioactive material is found in nature, and as a result of the development and use of atomic energy as a power source, and the mining of radioactive and other materials. It has become necessary to set upper limits for the intake of radioactive substances into the body, including intake from drinking water.

The effects of human exposure to radiation or radioactive materials are harmful and any unnecessary exposure should be avoided. The levels of radioactive materials specified in the current drinking water regulations (see Table 1) are intended to limit the human intake of these substances so that the radiation exposure of any individual will not exceed the amount defined by current radiation protection guidelines. People have always been exposed to natural radiation from water, food, and air. The total amount of radiation to which a person is exposed varies with the amount of this background, or natural, radioactivity as well as other factors, including exposure to man-made sources of radiation.

Natural Radiation. Natural radiation occurs primarily in ground water and includes radon, radium, and uranium. Radon is the most common radiological threat. The health risk from these contaminants comes mainly from the deposit of radium and uranium within the bones and, inhalation of air which has been contaminated by water or soil gas containing radon. Radon is known to cause lung cancer. Samples of source water should be sent to a qualified laboratory for analysis if natural radioactivity is suspected.

Man-Made Radiation. The health risks associated with man-made radiation are similar to the health risks associated with natural sources of radiation. Of the radioactive substances that increase the risk of cancer, tritium, strontium, cesium and others have been found in drinking waters sources.

Radiological data indicating radioactive substances in an area's drinking water may be available in publications of the U.S. Environmental Protection Agency, U.S. Public Health Service, U.S. Geological Survey, or from Federal, State, or local agencies. For information or recommendations on specific problems, the appropriate agency should be contacted.

SAFE DRINKING WATER ACT (SDWA)

Background

As mentioned in the Introduction, in 1974 the SDWA gave the federal government the responsibility to establish national drinking water regulations to protect public health. This act was further strengthened and expanded in 1986 by Congress, requiring EPA to regulate additional contaminants, define maximum contaminant levels, establish compliance deadlines and filtration guidelines, require disinfection and wellhead protection and strengthen enforcement.

Chemical, physical, radiological and bacteriological substances in drinking water which pose a health risk to the public are regulated by the Environmental Protection Agency under the SDWA. EPA establishes maximum contaminant levels (MCLs) for drinking water contaminants. If analysis of a water supply shows that any of the MCLs listed in Table 1 are exceeded, the water should not be used without treatment to remove the contaminant(s).

Not all substances are regulated because of their health risks. Secondary maximum contaminant levels (SMCLs) are established by EPA to assure an aesthetically acceptable water supply (see Table 2). Exceeding SMCLs may create taste, odor or appearance problems, and cause people to question the safety of their drinking water even though there is no direct health risk. As a result, people might use a water supply that looks, smells or tastes better, but poses a health risk.

TABLE 1. - *Maximum Contaminant Levels (MCLs)*[1]

Inorganic Contaminants

PARAMETER	MCL	PARAMETER	MCL
Arsenic	0.05	Fluoride	4.0
Asbestos (fibers/L)	7 million	Lead	0.015[2]
Barium	1	Mercury	0.002
Cadmium	0.005	Nitrate (as N)	10
Chromium	0.1	Nitrate + Nitrite (as N)	1
Copper	1.3[2]	Selenium	0.01
Fluoride	4.0		

Organic Contaminants

PARAMETER	MCL	PARAMETER	MCL
Alachlor	0.002	Ethylene dibromide	0.00005
Atrazine	0.003	Heptachlor	0.0004
Benzene	0.005	Heptachlor epoxide	0.0002
Carbofuran	0.04	Lindane	0.0002
Carbon Tetrachloride	0.005	Methoxychlor	0.04
Chlordane	0.002	Monochlorobenzene	0.1
2,4-Dichlorophenoxyacetic	0.07	Polychlorinated biphenyls	0.0005
Dibromochloropropane	0.0002	Styrene	0.1
o-Dichlorobenzene	0.6	Tetrachloroethylene	0.005
para-Dichlorobenzene	0.075	Toluene	1
1,2-Dichloroethane	0.005	Toxaphene	0.003
1,1-Dichloroethylene	0.007	Total Trihalomethanes	0.10
cis-1,2-Dichloroethylene	0.07	2,4,5-TP (Silvex)	0.05
trans-1,2-Dichloroethylene	0.1	Trichloroethane	0.005
1,2-Dichloropropane	0.005	1,1,1-Trichloroethane	0.20
Endrin	0.0002	Vinyl Chloride	0.002
Ethylbenzene	0.7	Xylenes	10

Radiological Contaminants

PARAMETER	MCL
Total radium (radium-226 + radium-228) (pCi/L)	5
Gross alpha activity (pCi/L)	15
Beta partical and photon activity (millirem/year)	4

1 Units in mg/L unless otherwise noted.

2 USEPA has set "action levels" for copper and lead in public drinking water.

Appendix A is a list of drinking water contaminants with associated health effects, sources of contamination, and best treatment methods also presented.

TABLE 2. - *Secondary Maximum Contaminant Levels (SMCLs)*[1]

PARAMETER	SMCL	PARAMETER	MCL
Aluminum	0.05 to 0.2	Manganese	0.05
Chloride	250	Odor (TON)	3
Color (PtCo units)	15	pH	6.5 to 8.5
Copper	1.0	Silver	0.1
Corrosivity	Non-corrosive	Sulfate	250
Fluoride	2.0	Total dissolved solids	500
Foaming Agents	0.5	Zinc	5.0
Iron	0.3		

1 Units in mg/L unless otherwise noted.

Maximum Contaminant Levels

The Safe Drinking Water Act requires EPA to set numerical standards, referred to as Maximum Contaminant Levels (MCLs), or define treatment requirements for 83 contaminants that pose potential health risks in drinking water. The 1986 amendments established a strict schedule for EPA to set MCLs or treatment requirements for many additional contaminants. Every three years, an additional 25 contaminants must be regulated. In addition to the MCLs and treatment techniques, secondary maximum contaminant levels (SMCLs) which effect the aesthetic quality of drinking water (but do not represent significant health hazards) have also been developed by EPA.

Monitoring

The EPA issues regulations that require regulated and some unregulated contaminants to be monitored. The unregulated contaminants that must be monitored depends on the number of people served, the water supply source, and contaminants likely to be found. Proper monitoring of biological, physical and chemical substances in water is necessary to insure a consistent, high quality supply of drinking water.

Filtration

Under the Safe Drinking Water Act, the EPA has set requirements for filtering surface water or water from sources that are exposed to surface water. It has also developed procedures for states to use in determining which systems have to filter their water.

Disinfection

The EPA must develop rules that require all public water supplies to disinfect their water. EPA rules will apply to water systems which use surface water sources and those which use groundwater.

Lead Levels

Solder or flux containing more than 0.2 percent lead, or pipes and pipe fittings containing more than 8 percent lead, cannot be used in new public water supply systems. Each public water system must identify and provide notice to persons whose drinking water may be contaminated due to lead in construction materials of the public water supply system, or where the corrosivity of the water may cause leaching of lead (from household plumbing or the distribution system) into the drinking water.

Wellhead Protection

The 1986 Safe Drinking Water Act amendments require all states to develop protection programs for the areas surrounding wells. These programs are designed to protect public water supplies from contamination.

Total Coliform Rule

EPA issued new regulations regarding total coliform occurrence in June 1988 that became effective December 31, 1990. The regulations establish limits and monitoring guidelines for total coliforms in public water systems to reduce the risk of water-borne disease caused by microorganisms. In addition, systems may be allowed to test for the bacterium Escherichia coli (E.coli) instead of fecal coliforms, and must follow special procedures to reduce the interference of heterotrophic bacteria (bacteria which get their energy from organic material) with the total coliform analysis. To comply with the new regulations, no more than 5.0 percent of the samples collected during a month can contain total coliforms. There is an exception for systems that collect fewer than 40 samples per month: they may have one total coliform-positive sample per month.

How often monitoring must take place is based on the number of people the system serves. The larger the system, the more samples must be collected. If a system collects fewer than five samples per month, it must also have a sanitary survey (described in this section) every five years (or every ten years for non-community water systems that use only protected and disinfected ground water). The reason for the sanitary survey requirement is that total coliforms are not found evenly distributed in the system. EPA does not believe that a system owner will be able to ensure that its water is safe based on fewer than five samples per month.

Under the coliform rule, if a sample is total coliform-positive, the system owner must collect three to four more samples to determine the extent of the contamination. Systems that collect fewer than fivc samples per month must (with some exceptions) collect at least five samples during the next month it serves the water to the public. This way, the system owner will collect at least ten samples over a two-month period, a number that EPA believes is necessary to find out whether the system is contaminated.

A state, however, cannot disregard a total coliform-positive sample just because all repeat samples are total coliform-negative. Again, this is because coliforms are distributed unevenly in the distribution pipes. These negative results do not prove that the system is free of contamination.

The total coliform rule also requires public water systems to test total coliform-positive samples for either fecal coliforms or E. coli to find out if the water is contaminated with fresh sewage. The presence of coliforms may mean recent sewage contamination, and the presence of a serious health concern. (Fecal coliforms are a category of total coliforms which includes E. coli and strains of several other bacteria.) If a sample contains fecal coliforms or E. coli, the system may be in violation of the total coliform rule. Variances to the total coliform rule may be available where positive coliform results can be shown to be caused by non-fecal microorganisms.

Surface Water Treatment Requirements

EPA Surface Water Treatment Rule (SWTR) requires all public water systems that use any surface water, or ground water under the direct influence of surface water (as determined by the State), to disinfect drinking water. All such systems must also filter their water, unless the State decides that filtration is not needed. More information about the total coliform rule and surface water treatment requirements for public water supplies can be found by asking the State or local agency responsible for water supply.

Control of Lead

EPA regulations include an "action level" for lead in household water, at 0.015 mg/L. Water systems that have lead levels above this action level in 10 percent or more of monitored homes would have to take several steps to minimize the exposure of people to lead in drinking water. These steps may include: installation or improvement of corrosion control; monitoring and possible treatment of source water supplies; delivery of a public education program to tell consumers ways that they can avoid high lead levels; and, in cases where corrosion control is not sufficient, gradual replacement of lead service connections that contribute more than 0.015 mg/L of lead to drinking water. Water systems are encouraged to help insure that plumbers in their locality comply with the 1986 Congressional ban on the use of solder containing greater than 0.2 percent lead and restrictions that limit to 8.0 percent the lead content of faucets, pipes and other plumbing materials.

WATER TESTING AND LABS

Many tests and analyses can be made on water to determine quality, so it is important to carefully select those most needed. County and state health departments may help in selecting tests and locating testing laboratories. Water treatment equipment dealers may also help determine appropriate tests, and may perform some analysis. Some states certify laboratories that are qualified to test water, while some states provide water testing services themselves.

Test kits are also available for analysis at the water source or supply (field analysis) of many chemical and bacteriological characteristics. Field test kits are available (from several suppliers) for alkalinity, bacteria (iron and sulfur bacteria, slime organisms and coliform), fluoride, hardness, iron, lead, manganese, chlorine, nitrate, pH, and zinc. Test kits are a useful indicator, but can not always be relied on to prove the presence or absence of a contaminant. If a drinking water contaminant is identified during a field test, it may be desired that a more detailed analysis be performed by a qualified laboratory.

Collection of Samples. Water samples must be collected carefully to prevent accidental contamination by the collector. Samples collected to satisfy drinking water regulations must be analyzed in a laboratory certified by the State or EPA. More information about sample collection and analysis for microorganisms in water is found in Appendix B.

CONTAMINANT SOURCES

There are many ways a water source could become contaminated. An accurate identification of the source of contamination is vital to determining the safety of a drinking water. If the source of contamination can be located, it may then be eliminated. Unfortunately, if the contamination cannot be eliminated, the water supply may still have to be used (with appropriate treatment) if no other water source is available.

Physical

Physical contamination can result from surface runoff during periods of heavy rainfall, urban runoff of oils, dirts and highway salts from roadways and parking lots, or from the growth of taste- and odor-causing algae in stored water.

Chemical

Chemical contamination may result from mine drainage, landfill leachate (leakage), storage tank leaks, accidental spills of chemicals in transport, agricultural fertilizer runoff, or salt water intrusion. Excessive hardness may also be present from the slow trickle (percolation) of water through natural mineral deposits.

Biological

Biological contamination comes from municipal, agricultural, industrial and individual wastewater systems. Treatment plant discharges, septic tank leaks, cess pools, and applying undisinfected wastewater sludge to farmland all contribute bacteria and viruses to source water. Other biological contamination can come from landfill leachate, deer and other wildlife in the watershed, and farm animal feedlots and manure lagoons.

Radiological

Radiological contamination comes from natural geologic formations and soils, and by leaks from radioactive storage sites, industrial sites, medical wastes and nuclear test sites. Natural radiation may also leak into drinking water sources from active or abandoned uranium mines.

Appendix C contains a useful key to identify many of the contaminants in water discussed above by using the senses of feeling, smell, taste, sight and hearing.

QUANTITY OF WATER

With quality water supplies becoming more difficult to find and water demands increasing, this limited resource must be conserved. In order to select a suitable water supply source, the demand that will be placed on it must be known. The elements of water demand include the average daily water use and the peak rate of demand. In the process, the ability of the water source to meet demands during critical periods (when surface flows and ground water tables are low) must be determined. Stored water, that would meet demand during these critical periods, must also be taken into consideration.

The "peak demand" rates must be estimated in order to determine plumbing and pipe sizing, pressure losses, and storage requirements necessary to supply enough water during periods of peak water demand.

State or local agency requirements may dictate water supply (and component) capacities. Where such agency requirements do not exist, the following discussion of average and peak demands can be used to project water needs.

Average Daily Water Use

Many factors influence water use. For example, the fact that water under pressure is available encourages people to water lawns and gardens, wash automobiles, and perform many other activities at home and on the farm. Modern household appliances such as food waste disposers and automatic dishwashers contribute to a higher total water use and tend to increase peak demands. Since water requirements will influence all features of a water supply under development or improvement, they are very important in planning. Table 3 presents a summary of average water use as a guide in preparing estimates. Local adaptations must be made where necessary, such as in cases where limitations in water supply require active water conservation measures.

TABLE 3. - *Planning guide for water use.*

Types of Establishments	Gallons per day
Airports (per passenger)	3-5
Apartments, multiple family (per resident)	60
Bath houses (per bather)	10
Camps: Construction, semipermanant (per worker)	50
Day with no meals served (per camper)	15
Luxury (per camper)	100-150
Resorts, day and night, with limited plumbing (per camper)	50
Tourist with central bath and toilet facilities (per person)	35
Cottages with seasonal occupancy (per resident)	50
Courts, tourist with individual bath units (per person)	50
Clubs: Country (per resident member)	100
Country (per non-resident member)	25
Dwellings: Boardinghouse (per boarder)	50
Luxury (per person)	100-150
Multiple-family apartments (per resident)	40
Rooming houses (per resident)	60
Single family (per resident)	50-75
Estates (per resident)	100-150
Factories (gallons per shift)	15-30
Highway rest area (per person)	5
Hotels with private baths (2 persons per room)	60
Hotels without private baths (per person)	50
Institutions other than hospitals (per person)	75-125
Hospitals (per bed)	250-400
Laundries, self-serviced (gallons per washing, i.e.,per customer)	50
Livestock (per animal):	
Cattle, horse, mule, steer (drinking)	12
Dairy (drinking and servicing)	35
Goat, sheep (drinking)	2
Hog (drinking)	4
Motels with bath, toilet, and kitchen facilities (per bed space)	50
With bed and toilet (per bed space)	40
Parks: Overnight with flush toilets (per camper)	25
Trailers with individual bath units, no sewer connections (per trailer)	25
Trailers with individual baths, connected to sewer (per person)	50
Picnic: With bathhouses, showers, and flush toilets (per picnicker)	20
With toilet facilities only (gallons per picnicker)	10
Poultry: Chickens (per 100)	5-10
Turkeys (per 100)	10-18
Resturants with toilet facilities (per patron)	7-10
Without toilet facilities (per patron)	2½-3
With bars and cocktail lounge (additional quantity per patron)	2
Schools: Boarding (per pupil)	75-100
Day with cafeteria, gymnasiums, and showers (per pupil)	25
Day with cafeteria but no gymnasium or showers (per pupil)	20
Day without cafeteria, gymnasiums, or showers (per pupil)	15
Service stations (per vehicle)	10
Stores (per toilet room)	400
Swimming pools (per swimmer)	10
Theaters (per seat)	5
Workers: Construction (per person per shift)	50
Day (school or offices per person per shift)	15

WATER CONSERVATION

Enormous amounts of treated water are wasted in the U.S. every day. Careless user attitudes, leakage, and bad habits cost thousands of gallons each day in even the smallest public water supply system. The cost of losing water after expensive, complex treatment is a serious concern. To reduce this problem, a water system may need to establish an active conservation strategy beginning at the treatment plant and extending out to even the smallest end user. If water saving conservation efforts are not made to preserve and protect water supplies, legal restrictions and water rationing may someday be forced upon water systems and users.

The average U.S. family of four uses an estimated 300 gallons of water every day. This is water that has been pumped, treated, and distributed or stored for consumption or use. About 95 percent of this water, or 285 gallons, ends up as sewage: As much as 120 gallons is flushed down toilets, with the remainder going down household drains.

Depending on the type of plumbing fixtures and personal use habits, average household water use is similar to the ranges described in Table 4:

Table 4.- ***Average Household Water Use Activities***

Per Person Per Day (Indoor Use Only)

Use	Gallons	Gallons per day	% Daily
Toilet (per flush)	1.5 - 5	25	37
Faucets (per minute use)	3	15	21
Bath/Shower (per minute use)	5	15	22
Daily laundry (per load)	25	10	15
Cooking/Drinking	3	3	5
		TOTAL	100%

Conservation Methods

Water conservation can be achieved by structural means, economic methods, usage restrictions or legal means. The following sections summarize how each of these may help a small public water supply system conserve.

Structural Means. Water saving devices, accurate metering and leak detection are physical methods of reducing water usage and thereby conserving water.

Water saving devices include household retrofit devices (new equipment installed in existing plumbing) and originally installed fixtures such as low-flow showers, faucets and toilets. Retrofit devices, such as toilet tank inserts, faucet aerators, and shower flow restrictors can be installed in existing home plumbing. Both originally installed low-flow equipment and retrofit devices can significantly reduce home water use.

A system can also reduce water demand by measuring (metering) customer water use and charging water rates that discourage excessive use. This may also serve to help customers identify the existence of leaks. If all water in a home is turned off and the meter continues to register, a leak exists that can be traced and repaired.

With meters, actual use is determined and the customer billed accordingly. If significant reductions in demand are required, a price increase can heighten a consumer's awareness of the need to conserve water. The largest user also pays the higher bill. Customers of a flat rate system, however, cannot be billed according to use since quantity used is not measured.

Another effective way to conserve water is to detect and repair leaks in the water system. This controls loss of water that a small system has already paid to acquire, treat, and pressurize. When this water leaks from the system before it reaches the consumer, the system loses money and gains unnecessary expenses.

A water audit compares the total quantity of water produced with metered water consumption. If the total metered water usage is less than 85 percent of the total metered water production, a system-wide leak detection survey should be conducted and follow-up repair work performed, if necessary.

Economic Methods. A price strategy based on metered water use can significantly reduce the demand for water. Flat rate systems, however, can only depend on restrictions and public appeals to reduce water usage. During a drought, or other cause for conservation, a metered system can also charge an increased amount for increased water usage to discourage waste.

Usage Restrictions. Water use restrictions can be adopted by either metered or unmetered small systems. Overall reduction of water use may be necessary due to insufficient treatment plant production capacity, limited availability of supply, or for the purpose of reducing peak water demand periods and thereby maintaining pressure in the distribution system. Restrictions often reduce overall water use by restricting outside water use to certain days or certain hours during a day.

Legal Means. Legal means of restricting water use by state or local agencies may include building code modifications and regulations designed to reduce water demand. Public ordinances banning car washing and lawn sprinkling have also been enforced to save water.

Public Awareness

Regardless which approach is taken to conserve water, a small system must also educate and involve water users about why such conservation measures are being imposed. Customers may react negatively to conservation measures unless a good explanation is provided. Water customers deserve an explanation of why the measures are required, how to conserve water and save money by using water saving devices, and how to inspect for leaks. Such knowledge develops good water saving habits.

FIRE PROTECTION

Small public water supply systems should also provide sufficient water for fire protection. An adequate water system for fire protection also results in lower fire insurance rates. Although potable water production is the main system responsibility, fire protection requirements have an important influence on the design and operation of most distribution systems. Fire protection consists of both public protection, available directly from hydrants supplied by the public distribution system, and private protection, such as building sprinkler systems.

Fire Flow Requirements

This is the rate of flow needed for fire fighting purposes to successfully fight a fire and confine the fire to a small area. Determination of this flow depends on the size, construction, occupancy of the building and its proximity to other buildings. This flow is calculated for different areas of a locality served. A water system will also have household water demands while fires occur; therefore, an adequate system must be able to deliver the required fire flow for a specified duration with system demand at the maximum daily rate. The maximum daily consumption is defined by insurance underwriters as the greatest total amount of water used during any 24-hour period in the past three years.

An estimated fire flow requirement for a given fire area can be calculated with the following formula from National Fire Codes:

$$F = 18\,C\,(A)^{0.5}$$

where:

F = fire flow requirement in gpm

C = building construction coefficient: 1.5 for wood frame, 1.0 for ordinary construction, 0.8 for noncombustible, and 0.6 for fire resistant construction

A = the total floor area of all floors in square feet excluding basement

Regardless of the calculated value, fire flow requirements should not be less than 500 gpm or more than 12,000 gpm. In residential districts the required flow ranges from a minimum of 500 gpm to a maximum of 2,500 gpm.

The period of time necessary to deliver the required fire flow is an important factor in water supply design because it directly influences the size of the storage facilities needed. Required fire flow duration is summarized in the table below adapted from the AWWA manual on Water Utility Management:

Required Fire Flow - gpm	Required Duration - Hour
10,000 and greater	10
9,000 - 10,000	9
8,000 - 9,000	8
7,000 - 8,000	7
6,000 - 7,000	6
5,000 - 6,000	5
4,000 - 5,000	4
3,000 - 4,000	3
2,500 and less	2

A suggested minimum storage capacity would be the maximum day usage plus fire requirements, less the daily capacity of the water plant and system for the fire flow period.

Required Pressure

If the small system supplies water to pumpers to fight fires, 20 psi is a minimum required pressure to provide a positive pressure at the pumper suction inlet while minimizing the chance of a negative pressure developing in the distribution system. In small systems that do not use pumpers, actual pressure in the distribution system is used to fight fires. A pressure of at least 60 psi is needed for small towns, and at least 50 psi in areas where no building exceeds two stories.

Distribution System

Supply mains and secondary feeders should be of sufficient size to deliver fire flow and maximum consumption demands to all areas. They should be properly spaced and looped so that no single area is totally dependent on one main.

A small system distribution grid should not contain mains smaller than 6 inches in diameter with no single interlock section longer than 600 feet. If street arrangement or topography does not allow this arrangement, then 8-inch diameter minimum mains should be used.

Valves

Valves should be arranged in shutoff lengths of 500 to 800 feet and arranged so repairs or breaks do not require shutdown of a main artery.

Hydrants

All water for public fire protection must be delivered through hydrants; therefore, there must be a sufficient number of them strategically placed throughout a small system.

Hydrant spacing is dictated by the required fire flow since the capacity of a single hydrant is limited. A single hose stream is considered to be 250 gpm. The minimum building area served by a single hydrant is commonly accepted as 40,000 square feet with a maximum area served of 100,000 square feet. Thus minimum hydrant spacing should be 200 feet, and in no case to exceed 500 feet apart. Ordinarily, hydrants are located at street intersections where streams can be pointed in any direction.

The ability to satisfy these fire protection requirements, whether through source development or careful control by means of conservation measures, ultimately determines the adequacy of a selected water supply source.

SANITARY SURVEY

A sanitary survey of water sources is very important. For a new supply, the sanitary survey should be made along with the collection of initial hydrogeologic and other information regarding the source and its ability to meet existing and future needs. The sanitary survey should include the detection of all existing and potential health hazards and the assessment of their present and future importance. Experts trained and competent in public health engineering and the epidemiology of waterborne diseases should conduct the sanitary survey. In the case of existing surface water supplies, a sanitary survey is recommended every three to five years to control health hazards and maintain high water quality. A periodic sanitary survey is required for some public systems under the total coliform regulations.

The sanitary survey involves three phases: planning the survey, conducting the survey, and compiling the final report. Those phases are presented in the following sections.

Planning the Survey

Before conducting or scheduling a sanitary survey, past sanitary survey reports, water system plans, any and all sampling results, operating reports and engineering studies should be found and closely studied. This pre-survey review provides an opportunity to see how and where samples are collected and how field measurements are made. The results of the pre-survey review should include a list of items to check while in the field, a list of questions about the system, and the format of the survey.

Conducting the Survey

This is the most important part of the survey. It involves interviewing those in charge of managing the system, reviewing major parts of the system from the source to distribution, and investigating any problems that are identified. Water sampling, and testing of equipment and facilities, are all part of the survey. Records of any field tests or water quality monitoring should also be checked to make sure proper maintenance is being performed.

Source Evaluation. Based on field observations and discussions, source evaluation should include a description of the area, stream flow, land usage, degree of public access, soil and ground cover. Next, sources of contamination, both man-made and natural, should be identified by visiting the area and asking others about the watershed. Surface intakes, infiltration galleries, springs and catchment/cistern systems should be evaluated in terms of their construction and their ability to protect the water supply. The source should be evaluated for its quantity, quality, protection from contamination, protection from damage, and ease of testing. All pumps, pump houses and controls should be checked for operation, maintenance, good condition and safety hazards. In addition, all check valves, blow off valves, water meters and other parts should be checked to make sure they operate and are being properly maintained. Finally, the identification and control of any contaminant sources must be determined in relation to the impacts on water quality.

Treatment Evaluation. Depending on the source water quality, one or more of several water treatment processes may be in use. These processes may include oxidation, coagulation, sedimentation, filtration and disinfection. In addition, more advanced processes such as granular activated carbon (GAC) adsorption and membrane separation may be part of the treatment system.

Disinfection is one treatment method which is common to many water supplies for insuring that drinking water is free of bacterial contamination. Disinfection feed, dosage and contact time should be evaluated. Critical spare parts, including a back-up disinfection unit in case of failure, may be required to be on hand. Proper storage procedures for the disinfectants and other safety precautions should also be followed to insure a safe, properly maintained system.

In all cases, treatment evaluation should include investigation of all treatment processes, availability of materials and spare parts, monitoring of process effectiveness, record keeping, competence of personnel who operate and maintain the system, and other key factors.

Distribution System Evaluation. Water quality during storage and distribution must also be maintained. The distribution system should have a scheduled program for removing any sediments that accumulate. The distribution system should also be inspected regularly for any possible sources of contamination. All distribution system overflow lines, vents, drainlines or cleanout pipes should be turned downward and screened. Capacity should be determined, and should be high enough to ensure sufficient pressure and flow. The reservoir should also be able to be isolated from the system.

Cross connections should be identified within the system. Each cross connection should be controlled and, preferably, eliminated. Annual testing of backflow prevention devices is advisable to assure that proper pressures and flows are maintained. In addition, a maintenance program should be developed to check leakage, pressure, corrosion control, working valves and hydrants. More information on detection and prevention of cross connections is provided in Part V.

Management/Operation. The overall operation of a system should be evaluated for its ability to deliver high quality drinking water. User fees should be charged to customers and collected regularly. There should be enough personnel to operate and manage the system. Operation and maintenance records should be kept readily available. Facilities should be free from safety defects. Emergency plans should be developed and usable, and include enough tools, supplies and maintenance parts.

The information furnished by a complete sanitary survey is essential for the interpretation of bacteriological and chemical water quality data. This information should always accompany the laboratory findings. The following outline covers the essential factors that should be investigated or considered in a sanitary survey. Not all of the items apply to any one system and, in some cases, items not in the list might be more important to a particular water supply.

Ground Water Supplies

a. Character of local geology; slope of ground surface.

b. Nature of soil and underlying porous material; whether clay, sand, gravel, rock (especially porous limestone); coarseness of sand or gravel; thickness of water-bearing layer, depth to water table; location, log (the underground features found by drilling a well), and construction details of local wells, both in use and abandoned.

c. Slope of water table, preferably as determined from observational wells or as indicated by the slope of ground surface (which is an inexact way estimating water-table slope).

d. Size of drainage area likely to contribute water to the supply.

e. Nature, distance, and direction of local sources of pollution (individual septic tanks [density per square mile] and soil percolation rate).

f. Possibility of surface or drainage water entering the supply and wells becoming flooded; methods of protection.

g. Methods used for protecting the supply against pollution by wastewater collection and treatment facilities and industrial waste disposal sites.

h. Well construction:
   1. Total depth of well.
   2. Casing: diameter, wall thickness, material, and length from surface.
   3. Screen or perforations: diameter, material, construction, locations, and lengths.
   4. Formation seal: material (cement, sand, bentonite, etc.), depth intervals, annular thickness, and method of placement.

i. Protection of well at top: presence of sanitary well seal, casing height above ground, floor, or flood level, protection of well vent, protection of well from erosion and animals.

j. Pumphouse construction (floors, drains, etc.), capacity of pumps, and whether water goes to storage or distribution system.

k. Drawdown when pumps are in operation; recovery rate when off.

l. Availability of alternate (although unsafe) water source that would require treatment.
m. Disinfection: equipment, back-up disinfection, supervision, test kits, or other types of laboratory control.
n. Availability of back-up power source (e.g., diesel generator) to protect supply during power outages.
o. Adequacy of supply to meet system demands (safe yield).
p. Adequacy of treatment to provide reliable, high quality drinking water.

Surface-Water Supplies

a. Nature of surface geology: character of soils and rocks.
b. Character of vegetation, forests, cultivated and irrigated land, including salinity, effect on irrigation water, etc.
c. Population and sewered population per square mile of watershed area.
d. Methods of sewage disposal, whether by diversion from watershed or by treatment.
e. Character and efficiency of sewage-treatment works on watershed.
f. Proximity of sources of fecal pollution to intake of water supply.
g. Proximity, sources, and character of industrial wastes, oil field brine, acid mine waters, etc.
h. Adequacy of quantity of supply (safe yield).
i. For lake or reservoir supplies: wind direction and velocity data, drift of pollution, sunshine data, algal blooms, stratified lake or impoundment.
j. Character and quality of raw water: concentration of coliform organisms, algae, turbidity, color, objectionable mineral constituents.
k. Nominal period of detention in reservoir or storage basin.
l. Probable minimum time required for water to flow from sources of pollution to reservoir and through reservoir intake.
m. Shape of reservoir, with reference to possible currents of water, induced by wind or reservoir discharge, from inlet to water-supply intake that may cause short-circuiting (uneven flow) to occur.
n. Measures to protect the watershed from controlled fishing, boating, landing of airplanes, swimming, wading, ice cutting, animals on marginal shore areas and in or upon the water.
o. Watershed control through ownership or zoning for restricted uses.
p. Efficiency and constancy of maintaining a controlled watershed.
q. Treatment of water: kind and adequacy of equipment; duplication of parts; effectiveness of treatment; adequacy of supervision, operation, and testing; contact period after disinfection; free chlorine residuals.
r. Pumping facilities: pumphouse, pump capacity and standby units, storage facilities.
s. Presence of an unsafe alternative source that may need treatment or have cross-connections that are a danger to public health.

Reporting the Survey

A final report of the survey should be completed as soon as possible after conducting the survey. The report can be used by the water system owner for daily operation and long term planning. It should include the date of the survey, who was present during the survey, the findings of the survey, the recommended improvements to problems found, and the dates set to complete improvements. The state agency responsible for water supply may also require a copy of the sanitary survey reports.

Part II

Ground Water

ROCK FORMATIONS

The rocks that form the crust of the earth are divided into three classes:

1. *Igneous.* Rocks that are derived from the hot magma deep in the earth are igneous. They include granite and other coarse crystalline rocks, dense igneous rocks (found in dikes and sills), basalt and other lava rocks, cinders, tuff, and other loose volcanic materials.
2. *Sedimentary.* Rocks that consist of chemical particles and rock fragments deposited by water, ice, or wind are sedimentary. They include deposits of gravel, sand, silt, clay, and the hardened equivalents of these--conglomerate, sandstone, siltstone, shale, limestone, and deposits of gypsum and salt.
3. *Metamorphic.* Rocks that are made up of both igneous and sedimentary rocks and are formed at great depths by heat and pressure are metamorphic. They include gneiss, schist, quartzite, slate, and marble.

The pores, joints, and crevices of the rocks in the saturation zone are generally filled with water. Although the openings in these rocks are usually small, the total amount of water that can be stored in subsurface reservoirs of rock formations is large. Those subsurface reservoirs are known as aquifers. The aquifers that contain the most water are deposits of clean, coarse sand and gravel; coarse, porous sandstones; cavernous limestones; and broken lava rock. Some dense limestones, as well as most of the igneous and metamorphic rocks do not hold much water. Silts and clays are among the most dense formations. The openings in these materials are too small to hold water, and the formations cannot maintain large openings under pressure. Dense materials near the surface, with crevice-like openings and cracks, may yield small amounts of water.

GROUND WATER BASINS

In an undeveloped ground water basin, movement of water to lower basins, seepage from and to surface-water sources, and transpiration depend on how much water is already in the basin and the rate of recharge. During periods of heavy rainfall, recharge may exceed discharge. If so, the excess rainfall increases the amount of water in the ground water basin. As the water table or artesian pressure increases, the water pressure at the points of discharge becomes higher and outflows increase. Extended dry periods will cause water-table levels and artesian pressures to decline. In most undeveloped basins, the major fluctuations in storage are seasonal, with the mean annual elevation of water levels changing little between years. Thus, the average annual inflow to storage equals the average annual outflow--a quantity of water referred to as "the basin yield".

Proper development of a ground water source requires careful consideration of the hydrological and geological conditions in the area. In order to take full advantage of a water source for domestic use, the assistance of a qualified ground water engineer, ground water geologist, hydrologist, or contractor familiar with the construction of wells in the area should be sought. Information on the geology and hydrology of an area should be available in publications of the U.S. Geological Survey or from other federal and state agencies. The National Water Well Association[1] also offers assistance.

SANITARY QUALITY

When water seeps through overlying material to the water table, particles in suspension, including micro-organisms, may be removed. How much is removed depends on the thickness and character of the overlying material. Clay, or "hardpan", provides the most effective natural filter for ground water. Silt and sand also provide good filtration if fine enough and in thick enough layers. The bacterial quality of the water improves during storage in the aquifer because conditions are usually unfavorable for bacteria. Clarity alone does not guarantee that ground water is safe to drink; this can only be determined by laboratory testing.

Ground water found in unconsolidated formations (sand, clay, and gravel) are protected from sources of pollution by these same types of materials and is likely to be safer than water coming from consolidated formations (limestone, fractured rock, lava, etc.).

Where overlying materials provide limited filtration, better and more sanitary water can sometimes be obtained by drilling deeper. It should be recognized, however, that there are areas where it is not possible, because of the geology, to find water at greater depths. Much unnecessary drilling has been done in the mistaken belief that more and better quality water can always be obtained by drilling deeper.

In areas without central sewerage systems, human excreta are usually deposited in septic tanks, cesspools, or pit privies. Bacteria in the liquid discharge from such installations may enter shallow aquifers. Sewage discharges have been known to find their way directly into water-bearing formations from abandoned wells or soil-absorption systems. In such areas, the threat of contamination may be reduced by proper well construction and/or by locating wells farther from sources of contamination. It is advisable to locate wells so that the normal movement of ground water carries contaminants away from the well.

CHEMICAL AND PHYSICAL QUALITY

The mineral content of ground water reflects its movement through the minerals which make up the earth's crust. Generally, ground water in arid regions is harder and more mineralized than water in regions with high annual rainfall. Also, deeper aquifers are more likely to contain higher concentrations of minerals in solution because the water

[1] National Water Well Association, 6375 Riverside Drive, Dublin, Ohio 43017. Phone: (614) 761-1711.

has had more time--perhaps millions of years--to dissolve the mineral rocks. For any ground water region, there is a depth below which salty water, or brine, is almost certain to be found. This depth varies from one region to another.

Some substances found naturally in ground water, while not necessarily harmful, may impart a disagreeable taste or undesirable property to the water. Magnesium sulfate (Epsom salt), sodium sulfate (Glauber's salt), and sodium chloride (common table salt) are a few of these. Iron and manganese are commonly found in ground water. Interestingly, regular users of waters containing so-called "excessive" amounts of these common minerals become accustomed to the water and consider it tasty.

Concentrations of chlorides and nitrates that are high for a particular region may be indicators of sewage pollution or contamination from agricultural fertilizers. This is another reason why a chemical analysis of the water should be made regularly and these results interpreted by someone familiar with the area.

Temperature

The temperature of ground water remains nearly constant throughout the year. Water from very shallow sources (less than about 50 feet or 15 meters deep) may vary somewhat from one season to another, but water from deeper zones remains quite constant--its temperature being close to that of the average annual temperature at the surface. This is why water from a well may seem relatively warm in winter or cool in summer.

Contrary to popular belief, colder water is not obtained by drilling deeper. Beyond about 100 feet of depth, the temperature of ground water increases steadily at the rate of about 1°F for each 75 to 150 feet of depth. In volcanic regions this rate of increase may be much greater.

DISTANCES TO SOURCES OF CONTAMINATION

All ground water sources should be located a safe distance from sources of contamination. In cases where sources are severely limited, however, a ground water aquifer that might become contaminated may be considered for a water supply if treatment is provided. After a decision has been made to use a water source in an area, it is necessary to determine the direction of water movement and the distance the source should be from the origin of contamination. It should be noted that the direction of ground water flow does not always follow the slope of the land surface. A determination of a safe distance is based on specific local factors described in the section on "Sanitary Survey" in Part I of this manual.

Because many factors determine the "safe" distance between ground water and sources of pollution, it is impractical to set fixed distances. The table on the following page offers guidance on determining safe distances for wells. Where sufficient information is unavailable, the distance should be the maximum that economics, land ownership, geology, and topography permit. Each installation should be inspected by a person with sufficient training and experience to evaluate all of the factors involved.

Since safety of a ground water source depends primarily on considerations of good well construction and geology, these should be the primary factors in determining safe distances for different locations. The following criteria apply only to properly constructed wells as described in this manual. There is no "safe" distance for a poorly constructed well!

When a properly constructed well penetrates an unconsolidated formation (sand, clay, gravel) with good filtering properties, and when the aquifer itself is separated from sources of contamination by similar materials, experience has demonstrated that 50 feet (or 15 meters) between the well and the source of the potential contamination is adequate. Shorter distances should be accepted only after qualified state or local health agency officials have conducted a comprehensive sanitary survey proving that shorter distances are safe.

When wells must be constructed in consolidated formations (limestone, fractured rock, lava formations, etc.), extra care should always be taken when locating the well and when setting "safe" distances, since pollutants have been known to travel great distances in such formations. The owner should request assistance from the State or local health agency.

If individuals propose to install a properly constructed well in formations of unknown character, the State or U.S. Geological Survey and the State or local health agency should be consulted.

The following table is offered as a guide in determining safe distances for wells:

Formations	Minimum acceptable distance from well to source of contamination
Favorable (unconsolidated)	50 feet. Lesser distances only on health department approval following comprehensive sanitary survey of proposed site and immediate surroundings.
Unknown	50 feet. **Only** after comprehensive geological survey of the site and its surroundings has established, to the satisfaction of the health agency, that favorable formations do exist.
Poor (consolidated)	Safe distances can be established only following both the comprehensive geological and comprehensive sanitary surveys. These surveys also permit determining the direction in which a well may be located with respect to sources of contamination. In no case should the acceptable distance be less than 50 feet.

EVALUATING CONTAMINATION THREATS

There are many sources of well contamination. Understanding the risks of contamination and locating wells far enough from potential sources of contamination can not only provide a safe, reliable supply of drinking water, but reduce the cost of treatment. Evaluating the following factors is the minimum action necessary to determine if a threat to a well exists.

Contaminants

Animal and human feces and toxic chemical wastes from plastics, solvents, pesticides, paints, dyes, varnishes, ink and other organic contaminants can be serious health threats. Other serious threats to human health can be agricultural chemicals such as pesticides and fertilizers. Salts, detergents, and solvents can dissolve and react with the ground water and migrate great distances with it. Many of these compounds are not removed by natural filtration. Contaminants of special concern are lead, radionuclides, bacteria, viruses, and organic compounds.

Lead. Use of brass and other lead-containing materials in operating wells should be avoided. Corrosive well water can corrode plumbing in lead service lines and lead solders contribute to lead in drinking water. Ingestion of large amounts of lead may cause impaired blood formation, brain damage, increased blood pressure, premature birth, low birth weight and nervous system disorders. Lead contamination especially poses serious health threats to young children.

Radionuclides. Radionuclides include naturally occurring substances that emit radiation as they decay, primarily in systems using ground water. Radionuclides such as radium and uranium in drinking water may be ingested into the body and cause cancer of the bone and kidney. Radon gas, which is dissolved in some ground waters, can be easily released in household air, causing a greater risk of lung cancer.

Bacteria and Viruses and Other Pathogens. Microbiological contaminants may enter ground water and eventually drinking water systems and result in highly contagious diseases. Symptoms of these diseases include nausea, vomiting, diarrhea and abdominal discomfort.

Organic Compounds. Other man-made organic materials may migrate from active or abandoned industrial waste sites. Many of these organic compounds migrate through ground water and can cause serious health risks.

Underground Disposal

Cesspools, septic systems, dry wells, disposal and injection wells of non-hazardous waste and hazardous waste, landfills, agricultural drainage, dumps and leaching pits that are in direct contact with the aquifer also increase the likelihood of contamination. This can be directly, by a contaminant leaching into the aquifer and into a drinking water well, or indirectly by reducing the amount of soil filtering the contaminants and allowing them to travel faster and further distances through the ground. In areas where direct disposal of wastes into sink holes or cavities (limestone terrains) has occurred, the risk of contamination is also increased.

Limited Filtration

When materials surrounding the well and overlying the aquifer are too coarse (limestone with solution cavities, gravel, etc.) or too thin (less than two feet) to provide effective filtration, the risk of contamination is increased. Lack of a fine grain surface layer permits faster infiltration of storm water and the contaminants dissolved in it.

Aquifer

When the aquifer materials are too coarse to provide good filtration (dissolved limestone, fractured rock, etc.), contaminants entering aquifer may travel great distances very rapidly. Often the exact destination of contaminants may be difficult to predict without the use of tracers. In such cases it is important to know the direction of ground water flow, and the locations of possible contamination sources in order to locate wells "upstream" of the contamination or outside the contaminated flow path.

Flow Changes

If large volumes of liquid waste are being discharged or ponded and are reaching the aquifer, the slope of the water table and direction of ground water movement can be altered. Water withdrawal will also change the direction and increase the speed of ground water movement. Increased withdrawal of water from wells directly increases the speed of ground water flow and may, therefore, increase the likelihood of contamination.

Multiple Threats

More than one source of contamination (e.g., domestic sewage and industrial wastes) in an area increases the total pollution load in the aquifer system and increases the risk of contamination. Also, interactions from different contaminants may in some instances produce hazardous compounds that were not originally present, increasing the risk to public health.

Wellhead Protection Areas

A wellhead protection area provides a protective barrier for the area surrounding a well or well field supplying a water system. Ensuring protection of the well or wells from possible contamination requires considerable investigation of the surface and ground water flows around the wells and locating sources of potential contamination. A protection area around a well should be a minimum of 50 feet between the well and any form of potential contamination, and can help ensure a safe supply of drinking water. Properly sealing the well to prevent any pollutants or rain water from entering the well and or selecting an aquifer which has no contamination sources "upstream" of the well are also effective.

As stated previously, if one plans to install a properly constructed well in a formation of unknown character, the State, U.S. Geological Survey, or local health agency should be consulted. These agencies can also provide advice on local groundwater flows and areas which should be avoided due to the potential for contamination.

DEVELOPMENT OF GROUND WATER

Ground water development depends on the geological formations and hydrological characteristics of the water-bearing formation. The development of ground water falls into two main categories:

1. Development by wells
    a. Nonartesian or water table
    b. Artesian
2. Development from springs
    a. Gravity
    b. Artesian

Nonartesian Wells

Nonartesian wells are those that penetrate formations in which ground water is found under water-table conditions. Pumping the well lowers the water table near it, creating a pressure difference and causing water to flow toward the well.

Artesian Wells

Artesian wells are those in which the ground water is under pressure because it is confined beneath an impermeable layer of material below the recharge area of the aquifer. Intake or recharge areas are commonly at high-level surface outcrops of the formations. Ground water flow occurs from high-level outcrop areas to low-level outcrop areas--which are areas of natural discharge. It also flows toward areas where water levels are lowered artificially by pumping from wells. When the well water level is higher than the top of the aquifer, the well is said to be artesian. A well that yields water by artesian pressure at the ground surface is a "flowing" artesian well.

Gravity Springs

Gravity springs occur in places where water percolating through permeable material that overlays an impermeable stratum comes to the surface. They also occur where the land surface intersects the water table. This type of spring is particularly sensitive to seasonal changes in ground water storage and frequently dwindles or disappears during dry periods. Gravity springs are characteristically low-volume sources, but when properly developed, they make satisfactory individual water supply systems.

Artesian Springs

Artesian springs discharge from artesian aquifers. They may occur where the confining impermeable formation is ruptured by a fault or where the aquifer discharges to a lower area. The flow from these springs depends on differences in recharge, discharge elevations and on the size of the openings transmitting the water. Artesian springs are usually more dependable than gravity springs, but the former are particularly sensitive to other wells being developed in the same aquifer. As a consequence, artesian springs may be dried by pumping.

Seepage Springs

Seepage springs are those in which the water flows (or seeps) out of sand, gravel, or other material containing many small openings. This includes many large and small springs. Some large springs have extensive seepage areas and are usually marked by plant growth. The water of small seepage springs may be colored or carry an oily scum because of decomposition of organic matter or the presence of iron. Seepage springs may emerge along the top of an impermeable bed, but they occur more commonly where valleys are cut into the zone of saturation of water-bearing areas. These springs are generally free from harmful bacteria, but they are susceptible to contamination by surface runoff which collects in valleys or depressions.

Tubular Springs

Tubular springs issue from relatively large channels, such as the solution channels from caverns of limestone, soluble rocks or smaller channels that occur in glacial drift. When the water reaches the channels by percolation through sand or other fine-grained material, it is usually free from contamination. When the channels receive surface water directly or receive the indirect effluent of cesspools, privies, or septic tanks, the water is usually unsafe for consumption.

Fissure Springs

Fissure springs issue along bedding, joint, cleavage, or fault planes. Their distinguishing feature is a break in the rocks along which the water passes. Some of these springs discharge uncontaminated water from deep sources. Many thermal springs are of this type. Fissure springs, however, may discharge water which is contaminated by drainage close to the surface.

Pumping Effects

When a well is pumped, the water level near the well is lowered (see Figure 2a). This lowering or "drawdown" causes the water table or artesian pressure surface, depending on the type of aquifer, to take the shape of an inverted cone called a "cone of depression". This inverted cone, with the well at the lowest point, is measured in terms of the difference between the static water level and the pumping level. At increasing distances from the well, the water level increases until it meets the static water table. The distance from the well at which this occurs is called "the radius of influence". The radius

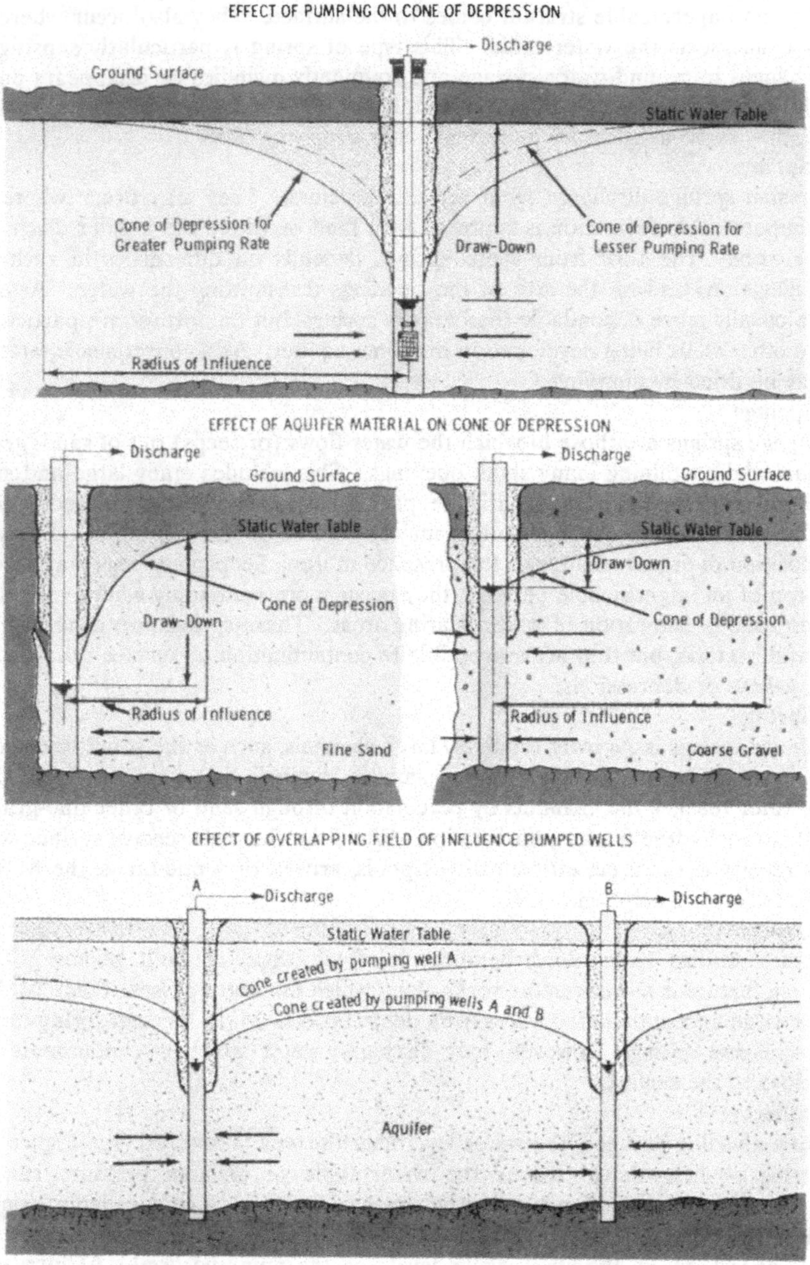

FIGURE 2. - ***Pumping effects on aquifers.***

of influence is not constant but tends to expand continuously with pumping. At a given pumping rate, the shape of the cone of depression depends on the characteristics of the water-bearing formation. Shallow and wide cones will form in highly permeable aquifers composed of coarse sands or gravel. Steeper and narrower cones will form in less permeable aquifers. As the pumping rate increases, the drawdown increases; consequently, the slope of the cone gets deeper and more narrow.

The character of the aquifer--artesian or water table--and its physical characteristics that affect the cone's shape include thickness, lateral extent, and the size and grading of sands or gravels. In a material of low permeability, such as fine sand or sandy clay, the drawdown will be greater and the radius of influence less than from very coarse gravel (see Figure 2b). For example, when other conditions are equal for two wells, expect that pumping costs for the same pumping rate to be higher for the well surrounded by material of lower permeability because of the greater drawdown.

When the cones of depression of the two or more wells overlap, the local water table will be lowered (see Figure 2c). This requires additional pumping lifts to obtain water from the interior portion of the group of wells. A wider distribution of the wells over the ground water basin will reduce the cost of pumping and will allow greater yield.

Yield of Wells

The amount of water that can be pumped from any well depends on the character of the aquifer and the construction of the well. Contrary to popular belief, doubling the diameter of a well increases its yield only about 10 percent; conversely, it decreases the drawdown only about 10 percent for the same pumping rate. The casing diameter should be chosen to provide space for proper installation of the pump. Individual wells seldom require casings larger than 6 inches, 4-inch wells are common.

A more effective way of increasing well capacity is by drilling deeper into the aquifer--provided that the aquifer is deep enough. Also, the inlet portion of the well (screen, perforations, slots) is important in determining the yield of a well in a sand or gravel formation. The amount of "open area" in the screened or perforated portion exposed to the aquifer is critical. Wells completed in consolidated formations are usually of open-hole construction; there is no casing in the aquifer.

It is sometimes difficult to accurately predict a well's yield before its completion. Studying the geology of the area and interpreting the results obtained from nearby wells gives some indication. This information will be helpful in selecting the location and type of well most likely to be successful. The information can also provide an indication of the yield to expect from a well.

A common way to describe a well's yield is to express its discharge capacity in relation to its drawdown. This relationship is called "the specific capacity" of the well and is expressed in "gallons per minute (gpm) per foot of drawdown." The specific capacity may range from less than 1 gpm per foot of drawdown for a poorly developed well--especially one in a "tight" aquifer--to more than 100 gpm per foot of drawdown for a properly developed well in a highly permeable aquifer.

Dug wells can be sunk only a few feet below the water table. This seriously limits drawdown during pumping, which in turn limits the yield of the well. A dug well that taps a highly permeable formation, like gravel, may yield 10 to 30 gpm or more with only 2 or 3 feet of drawdown. If the formation is primarily fine sand, the yield may be on the order of 2 to 10 gpm. These results refer to dug wells of common size.

Like dug wells, bored wells can also be sunk only a limited depth below the static water level--probably 5 to 10 feet into the water-bearing formation. If the well is nonartesian, the drawdown should not be more than 2 or 3 feet. If the well taps an artesian aquifer, however, the static water level will rise to some point above the aquifer. The available drawdown and the well's yield will then be increased. A bored well that taps a highly permeable aquifer and provides several feet of available drawdown may yield 20 gpm or more. If the aquifer has a low permeability or the water is shallow, the yield may be much lower.

Driven wells can be sunk to as much as 30 feet or more below the static water level. A well at this depth provides 20 feet or more of drawdown. The well's small diameter limits the type of pump that can be employed, so that under favorable conditions, the yield is limited to about 30 gpm. In fine sand or sandy clay formations of limited thickness, the yield may be less than 5 gpm.

Drilled and jetted wells can usually be sunk to such depths that the depth of the well's standing water and the available drawdown will vary from less than 10 to hundreds of feet. In productive formations of considerable thickness, yields of 300 gpm and more are readily attained. Drilled wells can be constructed in all instances where driven wells are used and in many areas where dug and bored wells are constructed. The larger diameter of a drilled well--compared with that of a driven well--permits the use of larger pumping equipment which can develop the aquifer's full capacity. Again, a well's yield varies greatly depending on the permeability and thickness of the formation, the construction of the well, and the available drawdown.

Table 5 provides details about penetrating various types of geologic formations by the methods indicated.

TABLE 5. - *Suitability of well construction methods to different geological conditions.*

Characteristics	Dug	Bored	Driven	Jetted
Range of practical depths	0-50 feet	0-100 feet	0-50 feet	0-100 feet
Diameter	3-20 feet	2-30 inches	1.25-2 inches	2-12 inches
Type of geologic formation:				
Clay	Yes	Yes	Yes	Yes
Silt	Yes	Yes	Yes	Yes
Sand	Yes	Yes	Yes	Yes
Gravel	Yes	Yes	Fine	1/4" pea gravel
Cemented Gravel	Yes	No	No	No
Boulders	Yes	Yes, if less than well diameter	No	No
Sandstone	Yes, if soft and/or fractured	Yes, if soft and/or fractured	Thin layers	No
Limestone			No	No
Dense igneous rock	No	No	No	No

Characteristics	Drilled		
	Percussion	Rotary	
		Hydraulic	Air
Range of practical depths	0-1,000 feet	0-1,000 feet	0-750 feet
Diameter	4-18 inches	4-24 inches	4-10 inches
Type of geologic formation:			
Clay	Yes	Yes	No
Silt	Yes	Yes	No
Sand	Yes	Yes	No
Gravel	Yes	Yes	No
Cemented Gravel	Yes	Yes	No
Boulders	Yes, when in firm bedding	(Difficult)	No
Sandstone	Yes	Yes	Yes
Limestone	Yes	Yes	Yes
Dense igneous rock	Yes	Yes	Yes

[1] Flow pressure is the pressure in the supply near the faucet or water outlet while the faucet or water outlet is wide open and flowing.

Preparation of Ground Surface at Well Site

A properly constructed well should exclude surface water from a ground water source to the same degree as the natural undisturbed geologic formation. The top of the well must be constructed so that no foreign matter or surface water enters. The well site should be properly drained and adequately protected against erosion, flooding, and damage or contamination from animals. Surface drainage should be diverted from the well.

CONSTRUCTION OF WELLS

Dug Wells

Constructed by hand, the dug well is usually shallow. It is more difficult to protect from contamination, although, if finished properly, may provide a satisfactory water supply. Because of advantages offered by other types of wells, dug wells should not be considered if it is possible to construct one of the other wells described later in this section.

Dug wells are usually excavated with pick and shovel. Workers can lift the excavated material to the surface with a bucket attached to a windlass or hoist. A power-operated clam shell or orange peel bucket may be used in holes greater than 3 feet in diameter where the material is principally gravel or sand. In dense clays or cemented materials, pneumatic hand tools are effective means of excavation.

To prevent the native material from caving, place "a crib" or lining in the excavation and move it downward as the pit deepens. The space between the lining and the undisturbed embankment should be backfilled with clean material. In the region of water-bearing formations, the backfilled material should be sand or gravel. Place cement grout to a depth of 10 feet below the ground surface to prevent entrance of surface water along the well lining.

Depending on the availability of materials and labor cost, dug wells may be lined with brick, stone or concrete. Pre-cast concrete pipe, available in a wide range of sizes, provides an excellent lining. This lining can be used as a crib as the pit deepens. When using the lining as a crib, a concrete pipe with tongue-and-groove joints and smooth exterior surface is preferred (see Figure 3).

Bell and spigot pipe may be used for a lining where it can be placed inside the crib or in an unsupported pit. This type of pipe requires careful backfilling to guarantee a tight well near the surface. The primary factor in preventing contaminated water from entering a dug well is the sealing of the well lining and preventing seepage of surface water at and near the well.

Most dug wells do not penetrate far below the water table because of the difficulty of manual excavation including handling cribs and linings. The depth of excavation can be increased by using pumps to lower the water level during construction. Because of their shallow penetration into the zone of saturation, many dug wells fail in times of drought when the water level recedes or when large quantities of water are pumped from the wells.

Bored Wells

Bored wells are commonly constructed with earth augers turned either by hand or by power equipment. Constructing such wells is feasible at depths of less than 100 feet provided the water requirement is low and the overlying material has non-caving properties and contains few large boulders. In suitable material, holes from 2 to 30 inches

Note:
Pump screen to be placed below point of maximum draw-down

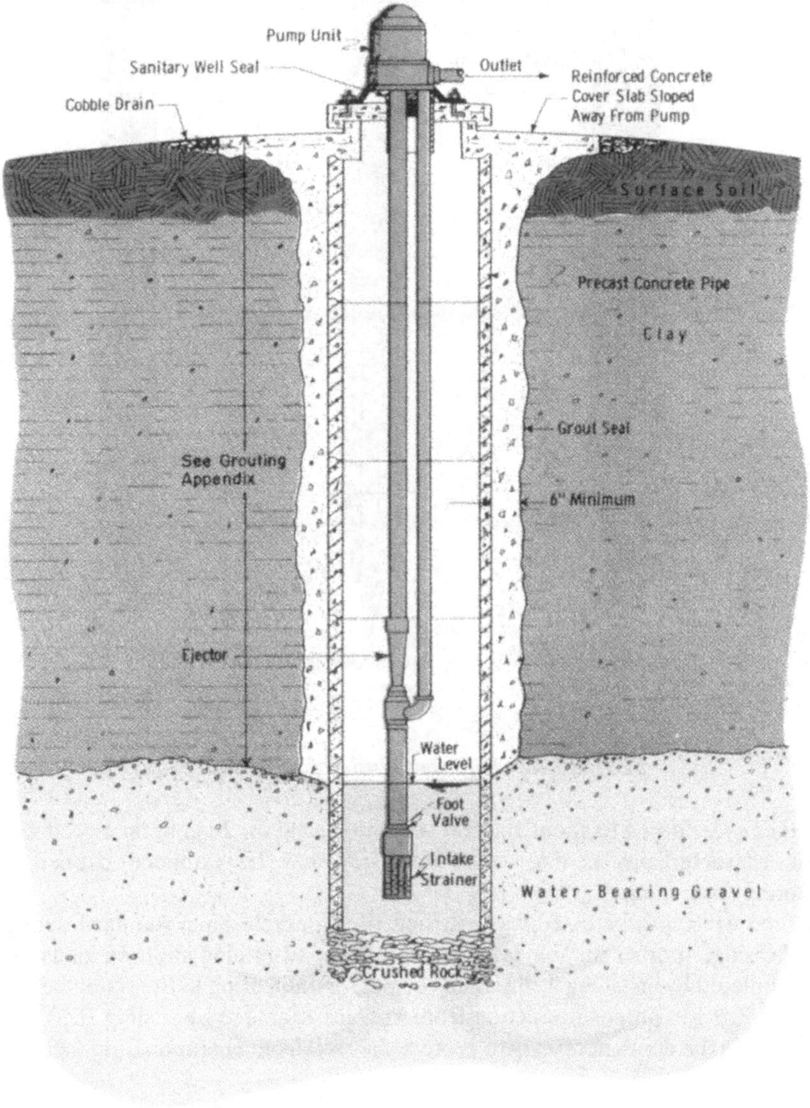

FIGURE 3. - ***Dug well with two-pipe jet pump installation.***

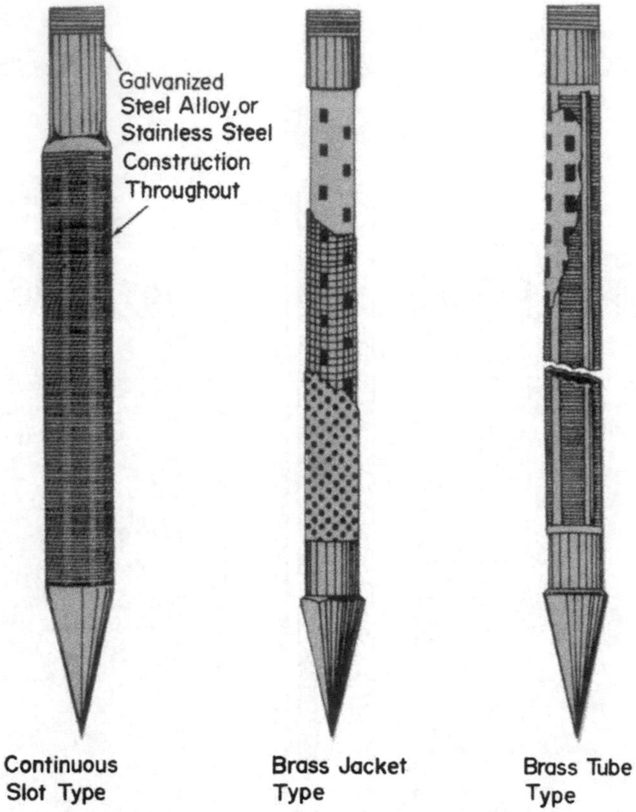

FIGURE 4. - ***Different kinds of drive-well points.***

in diameter can be bored to about 100 feet without caving in. In general, bored wells have the same characteristics as dug wells, but they may be extended deeper into the water-bearing formation.

Bored wells may be cased with vitrified tile, concrete pipe, standard wrought iron pipe, steel casing, or other suitable material capable of sustaining imposed loads. The well may be completed by installing well screens or perforated casing in the water-bearing sand and gravel. Provide proper protection from surface drainage by sealing the casing with cement grout to the depth necessary to protect the well from contamination (see Appendix C).

Driven Wells

The simplest and least expensive of all wells is the driven well. It is constructed by driving a drive-well point into the ground which is fitted to the end of a series of pipe sections (see Figures 4 and 5) made of forged or cast steel. Drive points are usually 1 1/4

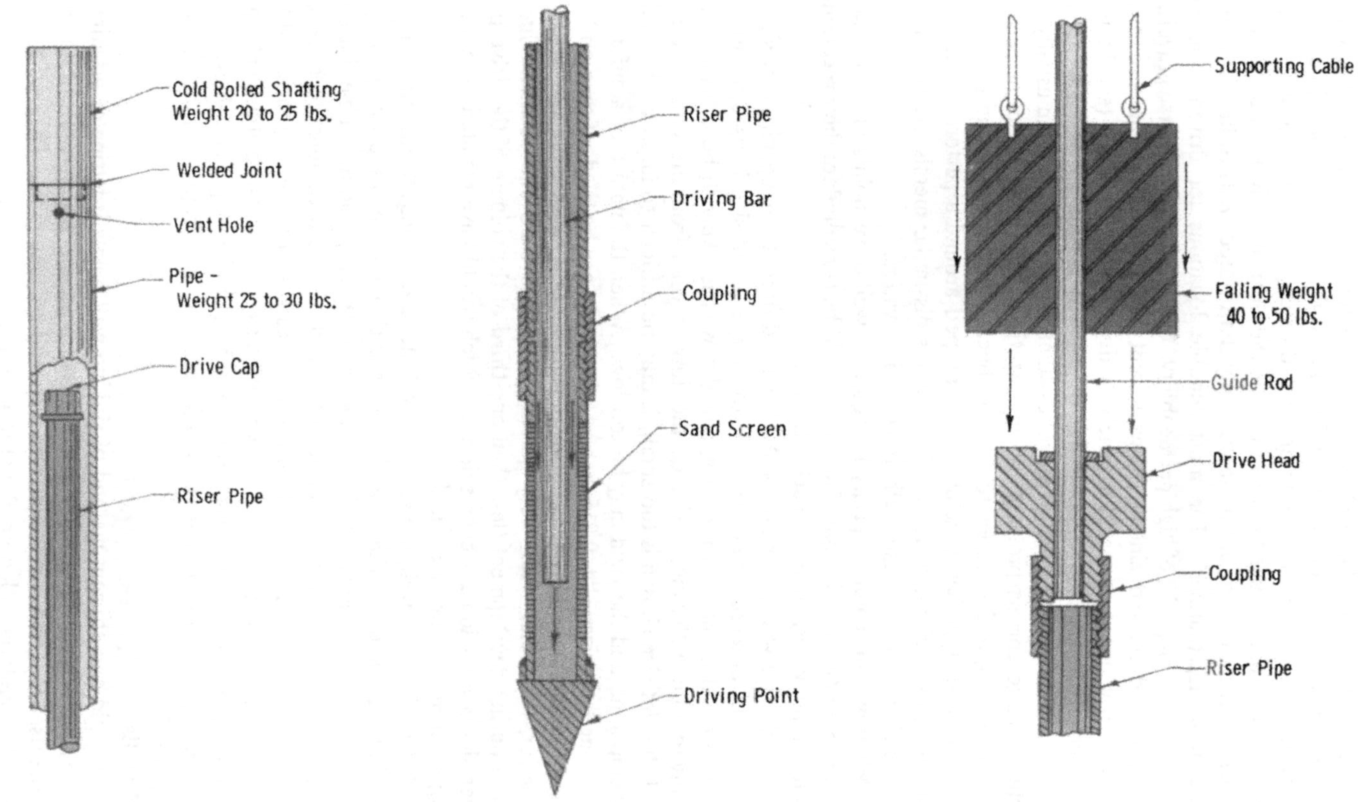

FIGURE 5. - ***Well-point driving methods.***

or 2 inches in diameter. The well is driven with the aid of a maul, or a special drive weight (see Figure 5). For deeper wells, the well points are sometimes driven into water-bearing strata from the bottom of a bored or dug well (see Figure 6). The yield of driven wells is usually small to moderate. If they can be driven an appreciable depth below the water table, they are no more likely than bored wells to be seriously affected by water-table fluctuations. The most suitable locations for driven wells are areas containing alluvial deposits of high permeability. The presence of coarse gravels, cobbles, or boulders interferes with sinking the well point and may damage the wire mesh jacket.

Well-drive points come in a variety of designs and materials (see Figure 6). In general, the efficiency and serviceability of each is related to its basic design. The continuous-slot, wire-wound type is more resistant to corrosion and can usually be treated with chemicals to correct problems of encrustation. It is more efficient because of its greater open area, and is easier to develop because its design permits easy access for cleanup. Another type has a metal gauze wrapped around a perforated steel pipe base and covered by a perforated jacket. If it contains dissimilar metals, electrolytic corrosion is likely to shorten its life--especially in corrosive waters.

Wherever maximum capacity is required, well-drive points of good design are a worthwhile investment. The manufacturers should be consulted for their recommendation of the metal alloy best suited to each situation.

Good drive-well points are available with different size openings, or slot sizes, for use in sands of different grain sizes. If too large a slot size is used, it may never be possible to develop the well properly, and the well is likely to be a "sand pumper," or gradually to fill in with sand, cutting off the flow of water from the aquifer. On the other hand, if the slot size chosen is too small, it may be difficult to improve the well capacity by development, and the yield may be too low. When the nature of aquifer sand is not known, a medium-sized slot--0.015 inch or 0.020 inch--can be tried. If sand and sediments continue to pass through the slots during development, a smaller slot size should be used. If, however, the water clears very quickly with little sand and sediment having been removed during development (less than one-third of the volume of the drive point) then a larger slot size could have been selected, which would have resulted in more complete development and greater well yield.

When driving a well, prepare a pilot hole that extends to the maximum practical depth. This can be done with a hand auger sightly larger than the well point. After the pilot hole has been prepared, lower the assembled point and pipe into the hole. Depending on the resistance of the formation, driving is accomplished in several ways. The pipe is driven by directly striking the drive cap, which is snugly threaded to the top of the protruding section of the pipe. A maul, a sledge, or a "special driver" may be used to hand-drive the pipe. The special driver may consist of a weight and sleeve arrangement that slides over the drive cap as the weight is lifted and dropped in the driving process (see Figure 5).

Jetted Wells

A rapid and efficient method of sinking well points is jetting or washing-in. This method requires a source of water and a pressure pump. Water is forced under pressure down the riser pipe and comes out from a special washing point. The well point and pipe is lowered as material is loosened by the jetting action of the water.

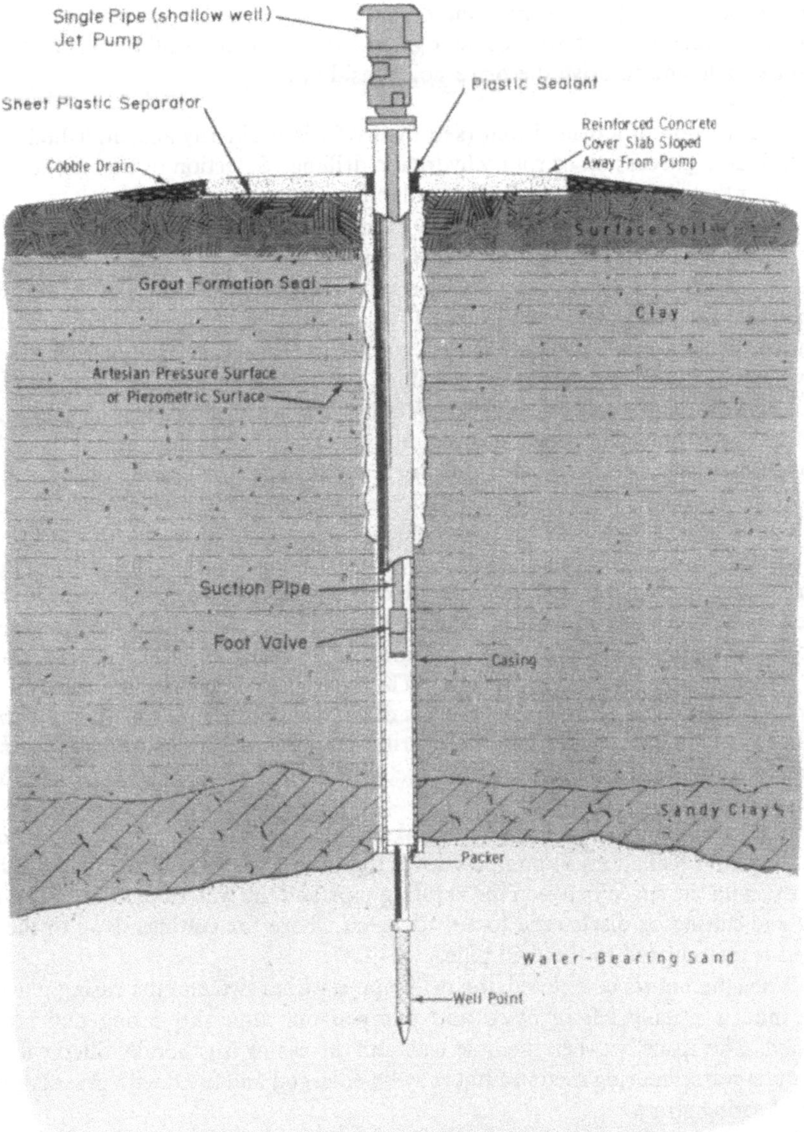

FIGURE 6. - ***Hand-bored well with driven-well point and "shallow well" jet pump.***

Often, the riser pipe of a jetted well is used as the suction pipe for the pump. In such instances, surface water may be drawn into the well if the pipe develops holes. An outside protective casing may be installed to the depth necessary to provide protection against the possible entry of contaminated surface water. The space between the two casings should then be filled with cement grout. It is best to install the protective casing in an auger hole and to drive the drive point inside it.

Drilled Wells

Construction of a drilled well (see Figure 7) is ordinarily accomplished by one of two techniques: percussion or rotary hydraulic drilling. Selection of the method depends primarily on the site's geology and the availability of equipment.

Percussion (Cable-Tool) Method. Drilling by the cable-tool or percussion method is done by raising and dropping a heavy drill bit and stem. The impact of the bit crushes and dislodges pieces of the formation. The reciprocating motion of the drill tools mixes the drill cuttings with water into a slurry at the bottom of the hole. This is periodically brought to the surface with a bailer, a 10- to 20-foot-long pipe equipped with a valve at the lower end. Caving is prevented as drilling progresses by driving or sinking a casing sightly larger in diameter than the bit. When wells are drilled in hard rock, casing is usually necessary only when drilling through layers of unconsolidated material. A casing may sometimes be necessary in hard rock formations to prevent caving of softer material.

When good drilling practices are followed, water-bearing beds are readily detected in cable-tool holes, because the slurry does not tend to seal off the water-bearing formation. A rise or fall in the water level in the hole during drilling, or increased recovery water during bailing, indicates that the well has entered a permeable bed. Crevices or soft streaks in hard formations are often water bearing. Sand, gravel, limestone, and sandstone are generally permeable and yield the most water.

Hydraulic Rotary Drilling Method. The hydraulic rotary drilling method may be used in most formations. The essential parts of the drilling assembly include a derrick and hoist, a revolving table through which the drill pipe passes, a series of drill-pipe sections, a cutting bit at the lower end of the drill pipe, a pump for circulation of drilling fluid, and a power source to drive the drill.

In the drilling operation, the bit breaks up the material as it rotates and advances. The drilling fluid (called mud) pumped down the drill pipe picks up the drill cuttings and carries them up the space between the rotating pipe and the wall of the hole. The mixture of mud and cuttings is discharged to a settling pit where the cuttings drop to the bottom and mud is recirculated to the drill pipe.

When the hole is completed, the drill pipe is withdrawn and the casing placed. The drilling mud is usually left in place and pumped out after the casing and screen are positioned. The space between the hole wall and the casing is generally filled with cement grout in non-water-bearing sections, but may be enlarged and filled with gravel at the level of water-bearing strata.

When little is known about the geology of the area, the search for water-bearing formations must be careful and deliberate--locating and testing all possible formations. Water-bearing formations may be difficult to recognize by the rotary method or may be plugged by the pressure of the mud.

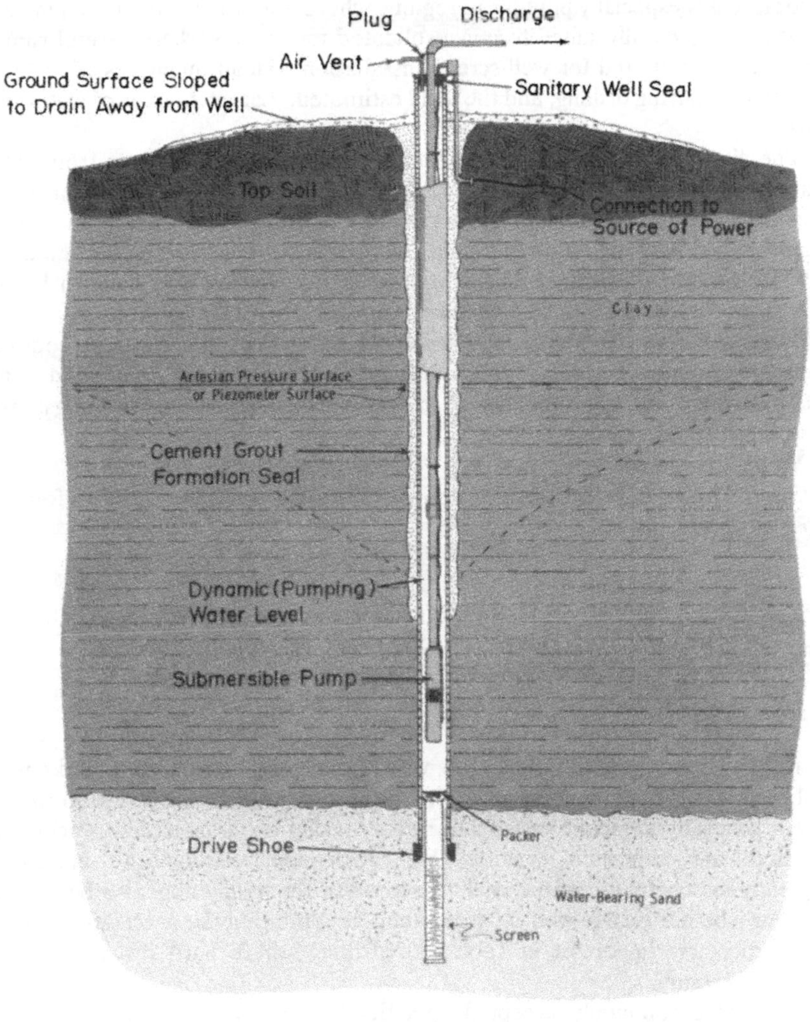

FIGURE 7. - *Drilled well with submersible pump.*

Air Rotary Drilling Method. The air rotary method is similar to the rotary hydraulic method in that the same type of drilling machine and tools may be used. The principal difference is that air is used rather than mud or water. In place of the conventional mud pump used to circulate the fluids, air compressors are employed.

The air rotary method is well adapted to rapid penetration of consolidated formations, and is especially popular in regions where limestone is the principal source of water. It is not generally suited to unconsolidated formations where careful sampling of rock materials is required for well-screen installation. Small quantities of water can be detected readily during drilling, and the yield estimated. Larger sources of water may slow progress.

The air rotary method requires that air be supplied at pressures from 100 to 250 pounds per square inch. To remove the cuttings, upward air velocities of at least 3,000 feet per minute are necessary. Penetration rates of 20 to 30 feet per hour in very hard rock are common with air rotary methods. Conventional mud drilling is sometimes used to drill through soft formations that overlie bedrock. Casing may have to be installed through those formations before continuing with the air rotary method.

Down-the-Hole Air Hammer. The down-hole pneumatic hammer combines the hammering effect of cable-tool drilling and the circular movement of rotary drilling. The tool bit is equipped with tungsten-carbide inserts at the cutting surfaces. Tungsten-carbide is very resistant to abrasion.

Water Well Casing and Pipe

Several kinds of steel pipe are suitable for casing drilled wells. The following are the most commonly used:

- Standard Pipe
- Line Pipe
- Drive Pipe
- Reamed and Drifted (R&D) Pipe
- Water Well Casing

For these there are different sizes, wall thicknesses, types of threaded connections available and methods of manufacture. The important thing for the owner to check is that they meet generally accepted specifications for quality of the steel and thickness of the wall. Both are important because they determine resistance to corrosion, and consequently the useful life of the well. Strength of the casing may also be important in determining whether certain well construction procedures may be successfully carried out. This is particularly important in cable-tool drilling where hard driving the casing is sometimes necessary.

The most commonly accepted specifications for water well casing are those prepared by the following:

- American Society for Testing and Materials (ASTM)
- American Petroleum Institute (API)
- American Iron and Steel Institute (AISI)
- Federal Government

Each source lists several specifications that might be used, but those most likely to be called for are ASTM A-120 and A-53, API 5-L, AISI Standard for R&D pipe, and Federal specification WW-P-406B. States may require certain well designs and/or materials, so the reader is advised to check with the appropriate agencies to determine the requirements.

Table 6 lists "standard weight" wall thicknesses for standard pipe and line pipe through the sizes ordinarily used in well construction. Thinner pipe should not be used. If conditions in the area are known to be highly corrosive, the "extra strong" and heavier weights should be used.

Setting Screens or Slotted Casings

Screens or slotted casings are installed in wells to permit sand-free water to flow into the well and to prevent unstable formations from caving in. The size of the slot for the screen or perforated pipe should be based on a sieve analysis of carefully selected samples of the water-bearing formation. The analysis is usually made by the screen manufacturer. If the slot size is too large, the well may pump sand. If the slots are too small, they may become plugged with fine material and reduce its yield. With a drilled well, the screens are normally placed after the casing has been installed. But in a driven well, the screen is a part of the drive assembly and is sunk to its final position as the well is driven.

TABLE 6. - ***Steel pipe and casing, standard and standard line pipe.***

Nominal size (inches)	Diameters		Wall thickness (inches)	Approximate weight (lb./ft.)	
	Outside	Inside		Plain ends	Threaded coupled
1-1/4	1.660	1.380	.140	2.27	2.30
1-1/2	1.900	1.610	.145	2.72	2.75
2	2.375	2.067	.154	3.65	3.75
3	3.500	3.068	.216	7.58	7.70
4	4.500	4.026	.237	10.79	11.00
5	5.563	5.047	.258	14.62	15.00
6	6.625	6.065	.280	18.97	19.45
8	8.625	8.071	.277	24.70	25.55
8	8.625	7.981	.322	28.55	29.35
10	10.750	10.192	.279	31.20	32.75
10	10.750	10.136	.307	34.24	35.75
10	10.750	10.020	.365	40.48	41.85
12	12.750	12.090	.330	43.77	45.45
12	12.750	12.000	.375	49.56	51.10

Consider the relationship between the open area of the screen and the velocity of water through the openings if maximum hydraulic efficiency is desired. Keep loss of energy through friction to a minimum by holding velocities to 0.1 foot per second or less. Since the slot size is determined by the grain-size distribution in the aquifer sand, the required open area must be obtained by varying the diameter--or, if aquifer thickness permits--the length of the screen. Manufacturers of well screens provide tables of capacities and other information to aid in selecting the most economical screen dimensions. Note that a screen is seldom required in wells tapping bedrock or tightly cemented sediments such as sandstone or limestone.

Methods for installing screens in drilled wells include (1) the pullback method, (2) the open-hole method, and (3) the baildown method. The pullback method of installation is one in which the casing is drawn back to expose a well screen placed inside the casing at the bottom of the well. In the open-hole installation, the screen attached to the casing is inserted in the un-cased bottom part of the hole when the aquifer portion of the hole remains open. When the baildown method is employed, the screen is placed at the bottom of the cased hole and advanced into the water-bearing formation by bailing the sand out from below the screen.

The "pullback" method is suited to bored or drilled wells, as long as the casing can be moved, while the "open-hole" method is used in most instances with rotary drilling. The "baildown" method may be used in wells drilled by any method where water-bearing formations consist of sand. It is not well adapted to gravel formations.

A fourth method, used primarily in rotary drilled holes, is the wash-down method. This procedure involves the circulation of water through a self-closing bottom, upward around the screen and through the space between the wash-pipe and the permanent casing to the surface. This is accomplished with the help of a mud pump. As material is washed by jet action from below it, the well screen settles to its desired position.

If the screen is placed after positioning the casing, it must be firmly sealed to the casing. This is done by spreading out a packer attached to the top of the screen. When the pullback method of installation is employed, a closed bail bottom usually provides the bottom closure. When the pullback method is used, a self-closing bottom serves this purpose. But when the baildown method is used, a special plug is placed in the bottom. A small bag of dry cement may be dumped into the bottom of the screen to seal it.

Development of Wells

Before a well is operational, it is necessary to remove all of the silt and fine sand next to the well screen by one of several processes known as "development". Development unplugs the formation and produces a natural filter of coarser and more uniform particles of high permeability surrounding the well screen. After the development is completed, a well-graded, stabilized layer of coarse material will entirely surround the well screen and facilitate the flow of water into the well.

The simplest method of well development is that of surging. In this process, the silt and sand grains are agitated by a series of rapid reversals in the direction of flow of water and are drawn toward the screen through larger pore openings. A well may be surged by moving a plunger up and down in it. This action moves the water alternately into and out of the formation. When water containing fine granular material flows into the well, the particles settle to the bottom of the screen. They then can be removed by pumping or bailing.

One of the most effective methods of development is the high-velocity hydraulic-jetting method. Water under pressure is ejected through the slot openings and violently agitates aquifer material. Sand grains finer than the slot size move through the screen and either settle to the bottom of the well (from which they are removed by bailing) or are washed out at the top. Conventional centrifugal, piston pumps, or the mud pump of the rotary hydraulic drill can easily accomplish this. Pressures of at least 100 psi should be used, with pressure greater than 150 psi preferred. This method has the additional advantage of permitting development of those portions of the screen most in need. For screens having continuous horizontal slot design, high-velocity jetting is most beneficial. It has also proven effective in washing out drilling mud and crevice cuttings in hard-rock wells. But it is less useful in slotted or perforated pipe.

Other methods of development include "interrupted pumping" and, explosives--used only in consolidated material and then only by experts. To insure proper well development, match the method of development to the aquifer and the type of well.

Testing Well for Yield and Drawdown

To select the most suitable pumping equipment, a pumping test should be made after the well has been developed to determine its yield and drawdown.
The pumping test for yield and drawdown should include the following:

- A determination of the volume of water pumped per minute or hour.
- The depth to the pumping level as determined over a period of time at one or more constant rates of pumping.
- The recovery of the water level after pumping is stopped.
- The length of time the well is pumped at each rate during the test procedure.

When the completed well is tested for yield and drawdown, it is essential that it be done accurately by using approved measuring devices and accepted methods. Additional information regarding the testing of wells for yield and drawdown may be obtained from the U.S. Geological Survey (USGS), the state or local health department, and the manufacturers of well screens and pumping equipment.

Water-table wells are more affected than artesian wells by seasonal fluctuations in ground water levels. When testing a water-table well for yield and drawdown, it is best, but not always possible, to test it near the end of the dry season. When this cannot be done, it is important to estimate the additional seasonal decline in water level from other wells tapping the same formations. This additional decline should then be added to the drawdown determined by the pumping test in order to arrive at the maximum pumping water level. Seasonal declines of several feet in water-table wells are not unusual, and can seriously reduce the capacity of such wells in the dry season.

Individual wells should be test pumped at a constant pumping rate that is not less than that planned for the final pump installation. The well should be pumped at this rate for not less than 4 hours, and the maximum drawdown recorded. Measurements of the water levels after pumping can then be made. Failure to recover completely to the original static water level within 24 hours is sufficient reason to question the dependability of the aquifer.

Well Failure

Over a period of time, wells may fail to produce for any of these causes:

1. Failure or wear of the pump.
2. Declining water levels.
3. Plugged or corroded screens.
4. Accumulation of sand or sediments in the well.

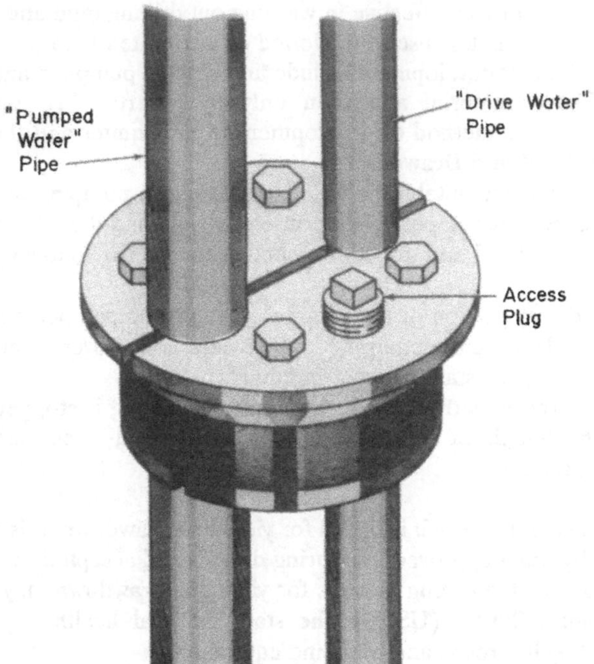

FIGURE 8. - *Well seal for jet pump installation.*

Proper analysis of the cause requires measuring the water level before, during, and after pumping. To facilitate measuring the water level, provide for the entrance of a tape or an electrical measuring device into the well in the space between the well casing and the pump column (Figures 7 and 8).

Use an "air line" with a water-depth indicating gauge (available from pump suppliers). On some larger wells, the air line and gauge are left so that periodic readings can be taken and a record kept of well and pump performance. While not as accurate as the tape or electrode method, this installation is popular for use in a well that is being pumped, because it is unaffected by water that may be falling from above.

Unless the well is the pitless adapter or pitless unit type, gain access for water-level measurements through a threaded hole in the sanitary well seal (Figures 8 and 9). This is possible for submersible and jet pump installations, as well as for some others. If it is not possible to gain access through the top of the well, do so by means of a pipe welded to the side of the casing.

If the well is completed as a pitless adapter installation, it is usually possible to slide the measuring device past the adapter assembly inside the casing and on to the water

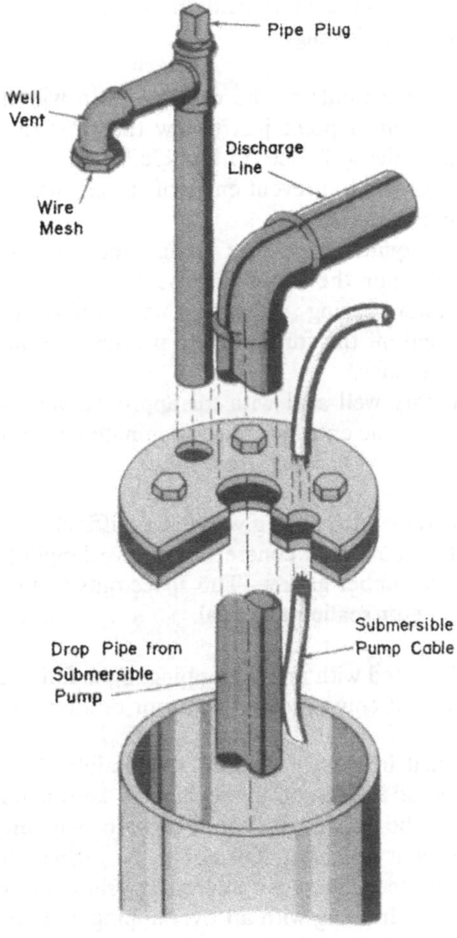

FIGURE 9. - *Well seal for submersible pump installation.*

below. If it is a pitless unit, particularly the "spool" type, it will probably not be possible to reach the water level. In the latter case, the well can only be tested by removing the spool and pump and reinstalling that pump, or another one, without the spool.

Any work performed within the well--including insertion of a measuring line--is likely to contaminate the water with coliform bacteria and other organisms. Before returning the well to service, it should be disinfected. Tightly plug or cover all access holes following the work.

SANITARY CONSTRUCTION OF WELLS

The penetration of a water-bearing formation by a well provides a direct route for possible contamination of the drinking water source. Although there are different types of wells and well construction, basic sanitary guidelines must be followed (state well codes may be more specific than the following):

1. Fill the open space outside the casing with a watertight cement grout or puddled clay from a point just below the frost line, or deepest level of excavation near the well (see "Pitless Units and Adapters," Page 128), to as deep as is necessary to prevent entry of surface water. See Appendix D for grouting recommendations.
2. For artesian aquifers, seal the casing into the overlying impermeable formations to retain the artesian pressure.
3. When a water-bearing formation containing poor-quality water is penetrated, seal off the formation to prevent the infiltration of water into the well and aquifer.
4. Install a sanitary well seal with an approved vent at the top of the well casing to prevent the entrance of contaminated water or other objectionable material.

For large-diameter wells, such as dug wells, it is difficult to provide a sanitary well seal. In such cases, install a reinforced concrete slab, overlapping the casing and sealed to it with a flexible sealant or rubber gasket. The space outside the casing should first be filled with a suitable grouting or sealing material.

Well Covers and Seals

Every well should be fitted with an overlapping, tight-fitting cover at the top of the casing or pipe sleeve to prevent contaminated water or other material from entering the well.

The sanitary well seal in a well exposed to possible flooding should be either watertight or elevated at least 2 feet above the highest known flood level. When it is expected that a well seal may be flooded, it should be watertight and equipped with a vent line whose opening to the air is at least 2 feet above the highest known flood level.

The seal in a well <u>not</u> exposed to possible flooding should be either watertight (with an approved vent line) or self-draining with an overlapping and downward flange. If the seal is of the self-draining (non-watertight) type, all openings in the cover should be either watertight or flanged upward and provided with overlapping, downward- flanged covers.

Some pump and power units have closed bases that effectively seal the upper end of the well casing. When the unit is the open type, or when it is located at the side (including some jet and suction-pump type installations), it is especially important that a sanitary well seal be used. There are several acceptable designs consisting of an expandable neoprene gasket compressed between two steel plates (see Figures 8 and 9). They are easily installed and removed for well servicing. Pump and well suppliers normally stock sanitary well seals.

If the pump is not installed immediately after well drilling and placement of the casing, the top of the casing should be closed with a metal cap screwed or tack-welded into place, or covered with a sanitary well seal.

A well slab alone is not an effective sanitary defense, since it can be undermined by burrowing animals and insects, cracked from settlement or frost heave, or broken by vehicles and vibrating machinery. The cement grout formation seal is far more effective. Note, however, that some situations call for a concrete slab or floor around the well casing to facilitate cleaning and improve appearance. When such a floor is necessary, it should be placed only after the formation seal and the pitless installation have been inspected.

Well covers and pump platforms should be elevated above the adjacent finished ground level. Construct pump room floors of reinforced, watertight concrete, that is carefully leveled or sloped away from the well so that surface and waste water will not stand near the well. The minimum thickness of such a slab or floor should be 4 inches. Concrete slabs or floors should be poured separately from the cement-formation seal. When there is the threat of freezing, insulate them from the well casing with a plastic or mastic coating or sleeve to prevent bonding of the concrete to either.

All water wells should be readily accessible at the top for inspection, servicing, and testing. This requires that any structure over the well be easy to remove, providing full, unobstructed access for well-servicing equipment. The so-called "buried seal," with the well cover buried under several feet of earth, is unacceptable because (1) it discourages periodic inspection and preventive maintenance, (2) it makes severe contamination during pump servicing and well repair more likely, (3) well servicing is more costly, and (4) excavation to expose the top of the well increases the risk of damage to the well, the cover, the vent and the electrical connections.

Disinfection of Wells

All newly constructed wells should be disinfected to neutralize any contamination from equipment, material, or surface drainage introduced during construction. Disinfection should be done promptly after construction or repair. State requirements may apply for disinfection of wells and need to be considered.

An effective and economical method of disinfecting wells and well equipment is the use of a calcium hypochlorite solution containing approximately 70 percent available chlorine. This chemical can be purchased in granular or tablet form at hardware stores, swimming pool equipment supply outlets, or chemical supply houses. When disinfecting wells, add calcium hypochlorite in sufficient amounts to provide a dosage of approximately 100 mg/L of available chlorine in the well water. This concentration is roughly equivalent to a mixture of 2 ounces of dry chemical per 100 gallons of water to be disinfected.

Practical disinfection requires the use of a stock solution. The stock solution may be prepared by mixing 2 ounces of high-test hypochlorite with 2 quarts of water. The solution should be prepared in a thoroughly clean utensil, avoiding metal containers if possible because strong chlorine solutions will corrode them. Crockery, glass, or rubber-lined containers are recommended. First add a small amount of water to the granular calcium hypochlorite and mix it to a smooth, watery lump-free paste. Then mix it with the remaining quantity of water. This stock solution should be mixed thoroughly for 10 to 15 minutes before allowing the inert ingredients to settle. The clearer liquid which contains the chlorine should be used and the sediment discarded. Each 2 quarts of stock solution provides a concentration of approximately 100 mg/L when added to 100 gallons of water.

When only small quantities of disinfectant are required and a scale is unavailable, measure dry chemicals with a spoon. A heaping tablespoonful of granular calcium hypochlorite weighs approximately 1/2 ounce.

When calcium hypochlorite is unavailable, other sources of available chlorine, such as sodium hypochlorite (12 to 15 percent of volume) can be used. Dilute an approved sodium hypochloride solution, available commercially with 5.25 available chlorine, with one part of water to produce a stock solution. Use two quarts of this solution for disinfecting 100 gallons of water.

Unless properly stored, stock solutions of chlorine in any form will deteriorate rapidly. Store in dark glass or plastic bottles with airtight caps. Keep bottles of the solution in a cool place and away from direct sunlight. If proper storage facilities are unavailable, always prepare fresh solution immediately before use. Because of its convenience and consistency of concentration and strength, commercial bleach is preferred as a stock solution for disinfecting individual water supplies.

Table 7 shows quantities of disinfectants required for treating wells of different diameters and water depths. For sizes or depths not shown, the next larger figure should be used.

Dug Wells. The disinfection procedure for dug wells is as follows:

1. After completing the casing or lining, follow the procedure outlined below before placing the cover platform over the well.
    a. Remove all equipment and materials, including tools, forms, platforms, etc., that will not be a permanent part of the completed structure.
    b. Using a stiff broom or brush, wash the interior wall of the casing or lining with a strong solution (100 mg/L of chlorine) to insure thorough cleaning.
2. Place the cover over the well and pour the required amount of chlorine solution as described above into the well through the manhole or pipe sleeve opening just before inserting the pump cylinder and drop-pipe assembly. Distribute the chlorine solution over as much surface area as possible to get the best distribution of the chemical in the water. Running the solution into the well through the water hose or pipeline while the line is being raised and lowered will insure better distribution.
3. Wash the exterior surface of the pump cylinder and drop pipe with the chlorine solution as the assembly is being lowered into the well.
4. After the pump is positioned, pump water from the well until a strong odor of chlorine is noted.
5. Allow the chlorine solution to remain in the well at least 24 hours.
6. After 24 hours or more have elapsed, flush the well to remove all traces of chlorine.

TABLE 7. - *Quantities[a] of calcium hypochlorite, 70 percent (rows A) and liquid calcium hypochlorite, 5.25 percent (rows B) required for water well disinfection.*

Depth of water in well (ft.)		Well diameter (in.)															
		2	3	4	5	6	8	10	12	16	20	24	28	32	36	42	48
5	A	1T	1T	1T	1T	1T	1T	2T	3T	5T	6T	3 oz.	4 oz.	5 oz.	7 oz.	9 oz.	12 oz.
	B	1C	1C	1C	1C	1C	1C	1C	1C	2C	4C	1Q	2Q	3Q	3Q	4Q	5Q
10	A	1T	1T	1T	1T	1T	2T	3T	5T	8T	4 oz.	6 oz.	8 oz.	10 oz.	13 oz.	1½ lb.	1½ lb.
	B	1C	1C	1C	1C	1C	1C	2C	2C	1Q	2Q	3Q	4Q	4Q	6Q	8Q	2½G
15	A	1T	1T	1T	1T	2T	3T	5T	8T	4 oz.	6 oz.	9 oz.	12 oz.	1 lb.	1½ lb.	1½ lb.	2 lb.
	B	1C	1C	1C	1C	1C	2C	3C	4C	2Q	2½Q	4Q	5Q	6Q	2G	3G	4G
20	A	1T	1T	1T	2T	3T	4T	6T	3 oz.	5 oz.	8 oz.						
	B	1C	1C	1C	1C	1C	2C	4C	1Q	2½Q	3½Q						
30	A	1T	1T	2T	3T	4T	6T	3 oz.	4 oz.	8 oz.	12 oz.						
	B	1C	1C	1C	1C	2C	4C	1½Q	2Q	4Q	5Q						
40	A	1T	1T	2T	4T	6T	8T	4 oz.	6 oz.	10 oz.	1 lb.						
	B	1C	1C	1C	2C	2C	1Q	2Q	2½Q	4½Q	7Q						
60	A	1T	2T	3T	5T	8T	4 oz.	6 oz.	9 oz.								
	B	1C	1C	2C	3C	4C	2Q	3Q	4Q								
80	A	1T	3T	4T	7T	9T	5 oz.	8 oz.	12 oz.								
	B	1C	1C	2C	4C	1Q	2Q	3½Q	5Q								
100	A	2T	3T	5T	8T	4 oz.	7 oz.	10 oz.	1 lb.								
	B	1C	2C	3C	1Q	1½Q	2½Q	4Q	6Q								
150	A	3T	5T	8T	4 oz.	6 oz.	10 oz.	1 lb.	1½ lb.								
	B	2C	2C	4C	2Q	2½Q	4Q	6Q	2½G								

[a]Quantities are indicated as: T = tablespoons; oz. = ounces (by weight); C = cups; lb. = pounds; Q = quarts; G = gallons.

NOTE: Figures corresponding to rows A are amounts of solid calcium hypochlorite required; those corresponding to rows B are amounts of liquid household bleach. For cases lying in shaded area, add 5 gallons of chlorinated water, as final step, to force solution into formation. For those in unshaded area, add 10 gallons of chlorinated water. (See "Disinfection of Wells," pp. 50 ff.)

Drilled, Driven, and Bored Wells. The procedure for disinfection of drilled, driven and bored wells are as follows:

1. When testing the well for yield, operate the test pump until the well water is as clear and as free from turbidity as possible.
2. After removing the testing equipment, slowly pour the required amount of chlorine solution into the well just before installing the permanent pumping equipment. Achieve best distribution of the solution with the well water by following Item 2 guidelines above, under "Dug Wells."
3. Add 5 or 10 gallons of clean, chlorinated water (see Table 7) to the well to force the solution out into the formation. One-half teaspoon of calcium hypochlorite or one-half cup of liquid sodium hypochlorite in 5 gallons of water is enough for this.
4. Wash the exterior surface of the pump cylinder and drop pipe as they are lowered into the well.
5. After positioning the pump, operate the pump until detecting a distinct odor of chlorine in the water.
6. Allow the chlorine solution to remain in the well for at least 24 hours.
7. After disinfection, pump the well until the odor of chlorine is gone.

In the case of deep wells having a high water level, it may be necessary to resort to special methods of introducing the disinfecting agent into the well to insure the best distribution of chlorine throughout the well.

Place the granulated calcium hypochlorite in a short section of pipe capped at both ends. Drill a few small holes through each cap or into the sides of the pipe. Fit one of the caps with an eye for attaching a suitable cable. The disinfecting agent is distributed when the pipe section is lowered or raised throughout the depth of the water.

Flowing Artesian Wells. The water from flowing artesian wells is generally free from contamination as soon as the well is completed or after it has been allowed to flow a short time. It is not generally necessary to disinfect flowing wells. If, however, analyses show persistent contamination, the well should be thoroughly disinfected according to the following guidelines.

Use a device such as the pipe described in the preceding section (or any other appropriate device) to distribute a generous supply of disinfectant at or near the bottom of the well. Pass the cable supporting the device through a stuffing box at the top of the well. After placing the disinfectant, throttle down the flow enough to get an adequate concentration. When water showing an adequate disinfectant concentration appears at the surface, close the valve completely and keep it closed for at least 24 hours.

Bacteriological Tests Following Disinfection

If bacteriological tests after disinfection indicates that the water is not safe for use, repeat disinfection until tests show that water samples from that portion of the system are satisfactory. Samples collected immediately after disinfection may not be representative of the normal water quality. Hence, if bacteriological samples are collected immediately after disinfection, it is necessary to re-sample several days later to check on the delivered water under normal conditions of operation and use. The water from the system should not be used for domestic purposes (such as drinking, cooking, brushing teeth, etc.) until the reports of the tests indicate that the water is safe for such uses. If after repeated disinfection the water is unsatisfactory, treatment of the supply is necessary to provide water which consistently meets USEPA drinking water requirements. Under these conditions, the supply should not be used for domestic purposes until adequate treatment has been provided.

ABANDONMENT OF WELLS

Unsealed, abandoned wells constitute a potential hazard to public health and safety. Certain problems with sealing an abandoned well may be due to the well's construction and the geological and hydrological conditions of the area. To seal a well properly, these main factors should be considered: elimination of any physical hazard, prevention of any contamination of the ground water, conservation and maintenance of the yield and hydrostatic pressure of the aquifer, and the prevention of any possible contact between acceptable and unacceptable waters.

The objective of properly sealing any abandoned well is to restore the controlling geological conditions existing before the well was drilled or constructed.

When a well is to be permanently abandoned, the lower portion of it is best protected by filling it with concrete, cement grout, neat cement, or clays having sealing properties similar to those of cement. When filling dug or bored wells, remove as much of the lining as possible, so that surface water will not reach the water-bearing layers through a porous lining or one containing cracks or fissures. When any question arises, follow the regulations and recommendations of the state or local health department.

Abandoned wells should never be used for the disposal of sewage or other wastes.

RECONSTRUCTION OF EXISTING DUG WELLS

Existing wells used for domestic water supplies and subject to contamination should be reconstructed to insure safe water. When reconstruction is not practicable, the water supply should be treated or a new well constructed.

Dug wells with stone or brick casings can often be rebuilt by enclosing existing casings with concrete or by installing a buried concrete slab.

Safety

Care must be exercised when entering wells because, until properly ventilated, they may contain dangerous gases or lack oxygen.

Hydrogen sulfide gas is found in certain ground waters and, being heavier than air, it tends to accumulate in excavations. It is explosive, and nearly as poisonous as cyanide. Also, a person's sense of smell tires quickly in its presence, making one unaware of the danger. Concentrations may become dangerous quickly without further warning.

Methane gas is also found in some ground water or in underground formations. It is the product of the decomposition of organic matter. It is not toxic, but is highly explosive.

Gasoline, carbide lanterns or candles may not be reliable indicators of safe well atmospheres because many of these devices will continue to burn at oxygen levels well below those safe for humans. Also, any open flame carries the additional risk of an explosion from accumulated combustible gases.

The flame safety lamp used by miners, construction companies, and utilities service departments, is a much better device for determining safe atmospheres. It is readily obtainable from mine safety equipment suppliers. The lamp should be lowered on a rope to the well bottom to test the atmosphere. Even after the well has passed this test, the first person to enter the well should carry a safety rope tied around his waist, with two persons standing by, above ground, to rescue him at the first sign of dizziness or other distress. Whenever possible, a self-contained air pack should be used by the person entering the well.

Improvements should be planned so that the reconstructed well will conform as nearly as possible to the principles set forth in this manual. If there is any doubt as to what should be done, advice should be obtained from the state or local health department.

SPECIAL CONSIDERATIONS IN CONSTRUCTING ARTESIAN WELLS

To conserve water and improve the productivity of an artesian well, it is essential that the casing be sealed into the confining stratum. Otherwise, a loss of water may occur by leakage into lower pressure permeable strata at higher elevations. A flowing artesian well should be designed so that the movement of water from the aquifer can be controlled. Equipping such a well with a valve or shutoff device conserves waters. When the recharge area and aquifer are large and the number of wells which penetrate the aquifer are small, the flowing artesian well produces a fairly steady flow of water throughout the year.

SPRINGS AND INFILTRATION GALLERIES

Springs and infiltration galleries on headwaters of a fresh water stream can be used to provide a safe, dependable source of drinking water. However, researchers have found disease-causing bacteria and protozoa in many of these water sources which would classify them as surface water sources, which are discussed in Part III of this manual.

Part III

Surface Water Source

INTRODUCTION

The selection and use of surface water as a water-supply source requires consideration of additional factors not usually related to ground water sources. For small public water supply systems, ground water or connection to a large central water system is generally preferred, and should be used whenever possible. When small streams, open ponds, lakes, or open reservoirs must be used, the danger of biological contamination and of the spread of diseases such as typhoid fever and dysentery increases. As a rule, *surface water should only be used when ground water sources are not available or are inadequate.*

Because surface water is open to physical and biological contamination, it is necessary to regard it as unsafe for household use unless reliable treatment, including filtration and disinfection, is provided.

EPA has set treatment requirements to control microbiological contaminants in public water systems using surface water sources (and ground water sources under the direct influence of surface water). These requirements, as part of the Surface Water Treatment Rule (SWTR), include the following:

- Treatment must remove or inactivate at least 99.9 percent of *Giardia lamblia* cysts and 99.99 percent of viruses.
- All systems must disinfect, and may also be required to filter if certain source water quality criteria and site-specific conditions are not met.
- The regulations set standards for determining if treatment, including turbidity removal and disinfection, is adequate for filtered systems.
- All systems must be operated by qualified, licensed operators as determined by the individual states.
- Systems using surface water must file reports with the state to show compliance with treatment and monitoring requirements.

Detailed guidance on surface water treatment requirements is provided in EPA's, *Guidance Manual for Compliance With the Filtration and Disinfection Requirements for Public Water Systems Using Surface Water Sources.*

Treatment of surface water to insure a consistent, safe supply requires regular attention to operation and maintenance of the treatment system by the owner. Public water systems must sample and monitor at a greater number of sampling sites. This includes regular monitoring of filter turbidities, disinfection residuals, and coliforms.

When ground water sources are limited, consideration should be given to reserving its use to household purposes such as drinking and cooking. Surface water can then be used for stock and poultry watering, gardening, firefighting, and similar purposes. Treatment of surface water for livestock is not generally considered essential. There is, however, a trend toward providing drinking water for stock and poultry which is free from bacterial contamination and certain chemicals.

SOURCES OF SURFACE WATER

Sources of surface water include rainwater catchments, ponds or lakes, and surface streams. The water in them comes from direct precipitation over the drainage area.

Because of the complexities of the hydrological, geological, and meteorological factors affecting surface water sources, it is recommended that engineering advice be obtained when developing a natural catchment area of more than a few acres.

To estimate the yield of the source, it is necessary to consider the following information for the drainage area:

- Total annual precipitation
- Seasonal differences of precipitation
- Annual or monthly variations of rainfall from normal levels
- Annual and monthly evaporation and transpiration rates
- Soil moisture requirements and infiltration rates
- Runoff gauge information
- Size of drainage area
- All available local experience records

Much of the required data, particularly that concerning precipitation, can be obtained from U.S. Weather Bureau publications. Essential data such as soil moisture and evaporation requirements may be obtained from local soil conservation and agricultural agencies or from field tests conducted by hydrologists.

Controlled Catchments

Historically, in many parts of the world, rainwater catchment has been used for water supply since ancient times. In areas where ground water is inaccessible or too highly mineralized for domestic use, controlled catchments and cisterns may be necessary. A properly located and constructed controlled catchment and cistern, with a good filtration unit and adequate disinfection facilities, will provide a safe drinking water.

A controlled catchment is a defined surface area from which rainfall runoff is collected. It may be a roof (roof catchment) or a paved ground surface (surface catchment). The collected water is stored in a constructed covered tank called a cistern or reservoir. Surface catchments should be fenced off to prevent unauthorized entrance by people or animals. There should be no possibility of undesired surface drainage mixing with the controlled runoff. An intercepting drainage ditch around the upper edge of the

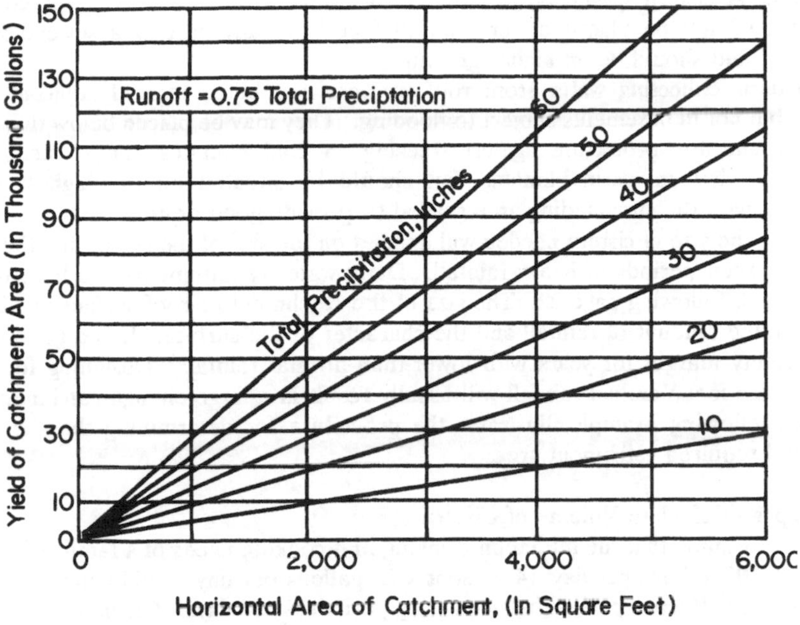

FIGURE 10. - ***Yield of impervious catchment area.***

catchment area and a raised curb around the surface will prevent the entry of any undesired surface drainage.

For these controlled catchments, simple guidelines to determine water yield from rainfall totals can be established. When the controlled catchment area has a smooth surface (or is paved) and the runoff collected in a cistern, water loss due to evaporation, replacement of soil moisture deficit, and infiltration is small. As a general rule, losses from smooth concrete or asphalt-covered ground catchments are less than 10 percent, shingled roofs or tar and gravel surfaces lose less than 15 percent, and sheet metal roofs lose almost no water.

A conservative design can be based on the assumption that the amount of water that can be recovered for use is three-fourths of the total annual rainfall. (See Figure 10)

Location. A controlled catchment may be located on a hillside near the edge of a natural bench. The catchment area can be placed on a moderate slope above the receiving cistern.

The location of the cistern should be governed by both convenience and quality protection. A cistern should be as close to the point of ultimate use as practical. A cistern should not be placed closer than 50 feet from any part of a sewage-disposal installation, and should be on higher ground.

Cisterns collecting water from roof surfaces should be located adjacent to the building, but not in basements subject to flooding. They may be placed below the surface of the ground for protection against freezing in cold climates and to keep water temperatures low in warm climates, but should be situated on the highest ground practicable, with the surrounding area graded to provide good drainage.

Size. The size of cistern needed will depend on the size of the family and the length of time between periods of heavy rainfall. Daily water requirements can be estimated from Table 3, following page 22. The size of the catchment or roof surface needed will depend on the amount of rainfall and the character of the surface. It is a good idea to allow a safety margin for years with lower-than-normal rainfall. Designing for three-fourths of the mean annual rainfall will usually result in a large enough catchment area.

The following example illustrates the procedure for determining the size of the cistern and required catchment area.

Step 1 - Calculate Volume of Cistern

Assume that the minimum drinking and cooking needs of a family of four are 100 gallons per day[1] (4 persons x 25 gallons per day = 100 gallons) and that the effective period[2] between rainy periods is 150 days. The minimum volume of the cistern required will be 15,000 gallons (100 x 150). This volume could be held by a cistern 10 feet deep and 15 feet square.

Step 2 - Calculate Required Catchment Area

If the mean annual rainfall is 50 inches, then the total design rainfall is 33 inches (50 x 2/3). In Figure 10, the catchment area required to produce 36,500 gallons (365 days x 100 gallons per day), the total year's requirement, is 2,400 square feet.

Construction. Cisterns should be water tight with smooth interior surfaces. Manhole or other covers should be sealed tight and vents screened to prevent the entrance of light, dust, surface water, insects, and animals.

Manhole openings should have a watertight curb with edges projecting a minimum of 4 inches above the level of the surrounding surface. The edges of the manhole cover should overlap the curb and project downward a minimum of 2 inches. The covers should be locked to minimize the danger of contamination and accidents.

---

1 Twenty-five gallons per person per day, assuming that other uses are supplied by water of poorer quality.

2 Effective period is the number of days between periods of rainfall during which there is negligible precipitation.

Provision can be made for diverting initial runoff, or "first flush", from paved surfaces or roof tops before allowing water to enter the cistern. Dirt, leaves, or bird droppings that accumulate on the roof or catchment area during dry periods should be washed down by the first flush of rain and collected in the roof washer. The drain at the lower end of the cistern is for cleaning, which may periodically be needed, because some contaminants may reach the cistern if the diversion of the first flush is not complete. (See Figure 11.)

Inlet, outlet, and waste pipes should be well-screened. Cistern drains and waste or sewer lines should not be connected.

Underground cisterns can be built of brick or stone, although reinforced concrete is preferred. If used, brick or stone must be low in permeability and laid with full portland cement mortar joints. Brick should be wet before laying. *High-quality workmanship is required, and the use of unskilled labor for laying brick or stone is not advised.* Two 1/2-inch plaster coats of 1:3 portland cement mortar on the interior surface will help to provide waterproofing. A hard, non-absorbing surface can be made by troweling the final coat before it is fully hardened.

Figure 11 shows a suggested design for a cistern of reinforced concrete. A dense concrete should be used to obtain water-tightness and should be vibrated frequently during construction to eliminate "honeycomb." All masonry cisterns should be allowed to wet-cure properly before being used.

The procedures outlined in Part IV of this manual should be followed in disinfecting the cistern with chlorine solutions. Initial and regular water samples should be taken to determine the bacteriological quality of the water supply. Chlorination may be required on a continuing basis if the bacteriological results show that the water quality is unsatisfactory. Roofs, gutters and other surfaces that collect and transport water for a cistern should be cleaned regularly. Roof gutters should be maintained at an even slope to prevent pooling of water. Roofs collecting rainwater should not be painted, nor should asbestos cement sheeting be placed so that it ever comes into contact with water. Tiles, slate, and galvanized iron are suitable for collection surfaces.

Ponds or Lakes

A pond or lake should be considered as a source of water supply only if ground water sources and controlled catchment systems are inadequate or unacceptable. The development of a pond as a supply source depends on several factors:

- A watershed that allows only water of the highest quality to enter the pond.
- Use of the best-quality water from the pond.
- Filtering the water to remove turbidity and reduce bacteria.
- Disinfection of filtered water.
- Proper storage of the treated water.
- Proper maintenance of the entire water system. Local authorities may be able to furnish advice on pond development.

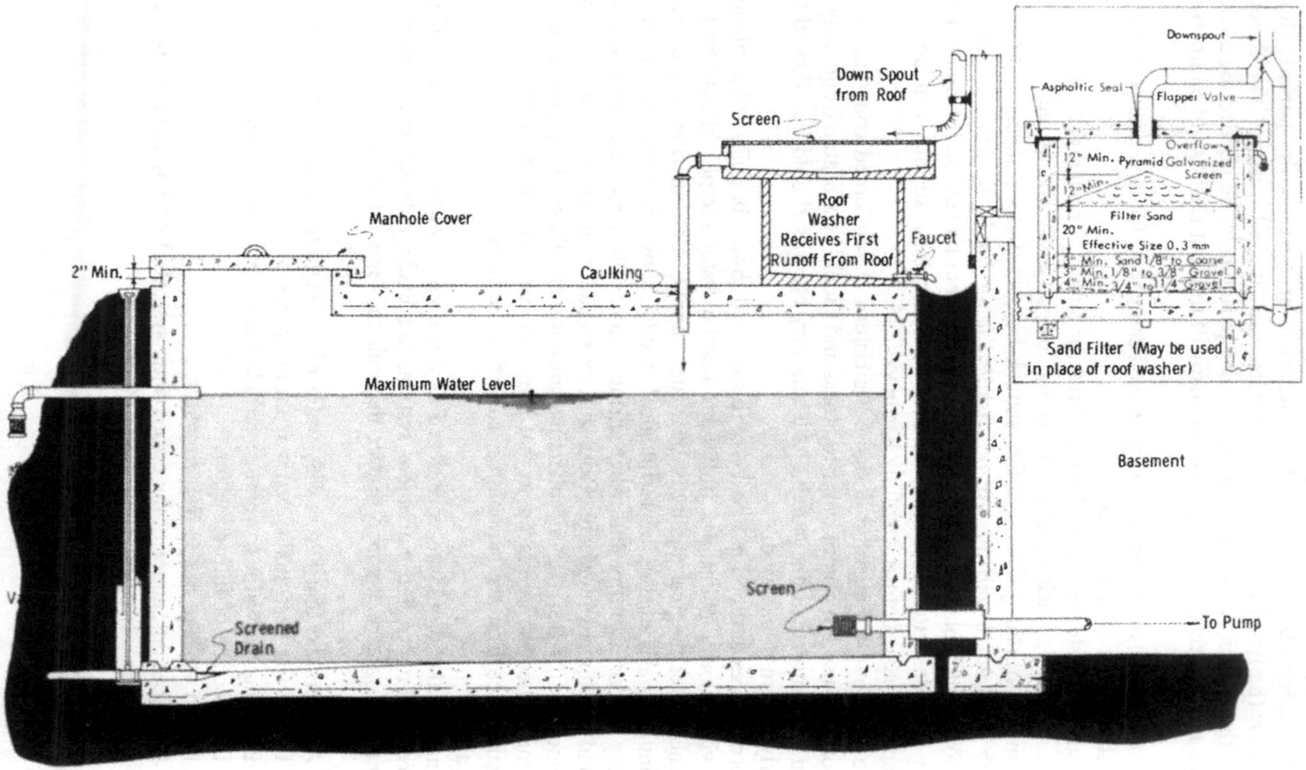

Down Spout
from Roof
Screen
Manhole Cover
2" Min.
Roof
Washer
Receives First
Runoff From Roof
Faucet
Caulking
Maximum Water Level
Screen
Screened
Drain
To Pump
Basement
Downspout
Asphaltic Seal
Flapper Valve
Overflow
12" Min.
Pyramid Galvanized
Screen
Filter Sand
20" Min.
Effective Size 0.3 mm
Sand Filter (May be used
in place of roof washer)

FIGURE 11. - *Cistern.*

The value of using a pond or lake as a water supply is its ability to store water during wet periods for use during dry periods. A pond should be capable of storing at least one year's supply of water. It must be big enough to meet water supply demand during periods of low rainfall, with an additional allowance for seepage and evaporation losses. The drainage area (watershed) should be large enough to fill the pond or lake during wet seasons of the year.

Careful consideration of the location of the watershed and pond site reduces the possibility of chance contamination.

The watershed should be:

- Clean, preferably grassed
- Free from barns, septic tanks, privies, and soil-absorption fields
- Effectively protected against erosion and drainage from livestock areas
- Fenced to keep out livestock

The pond should be:

- Not less than 8 feet deep at its deepest point.
- Designed to have the maximum possible water storage area that is over 3 feet in depth
- Large enough to store at least one year's supply
- Fenced to keep out livestock
- Kept free of weeds, algae, and floating debris

In many instances, pond development requires the construction of an embankment with an overflow or spillway. Help in designing a storage pond may be available from Federal, State, or local health agencies and in publications from the State or county agricultural, geological, or soil conservation departments. For specific conditions, professional engineering or geological advice may be needed as pond development entails potential liability that may be considerable.

Intake. A pond intake must be properly located so that it draws water of the highest possible quality. When the intake is placed too close to the pond bottom, it may draw cloudy water or water containing decayed organic material. When placed too near the pond surface, the intake system may draw floating debris, algae, and aquatic plants. The depth at which it operates best will vary, depending upon the season of the year and the layout of the pond. The most desirable water is usually obtained when the intake is located between 12 and 18 inches below the water surface. An intake located at the deepest point in the pond makes maximum use of stored water.

Pond intakes should be of the type illustrated in Figure 12, known as "floating intakes." The intake consists of a flexible pipe attached to a rigid pipe that passes through the pond embankment.

Gate valves should be installed on the main line below the dam and on any branch line to make it possible to control of the rate of discharge.

Treatment. The pond water-treatment facility consists of four basic parts. These are coagulation and settling for large particle removal, filtration for smaller particle removal, a clear water storage tank, and continuous disinfection (see Figure 13). A more detailed explanation of these treatment techniques is found in Part IV, under "Surface Water Treatment."

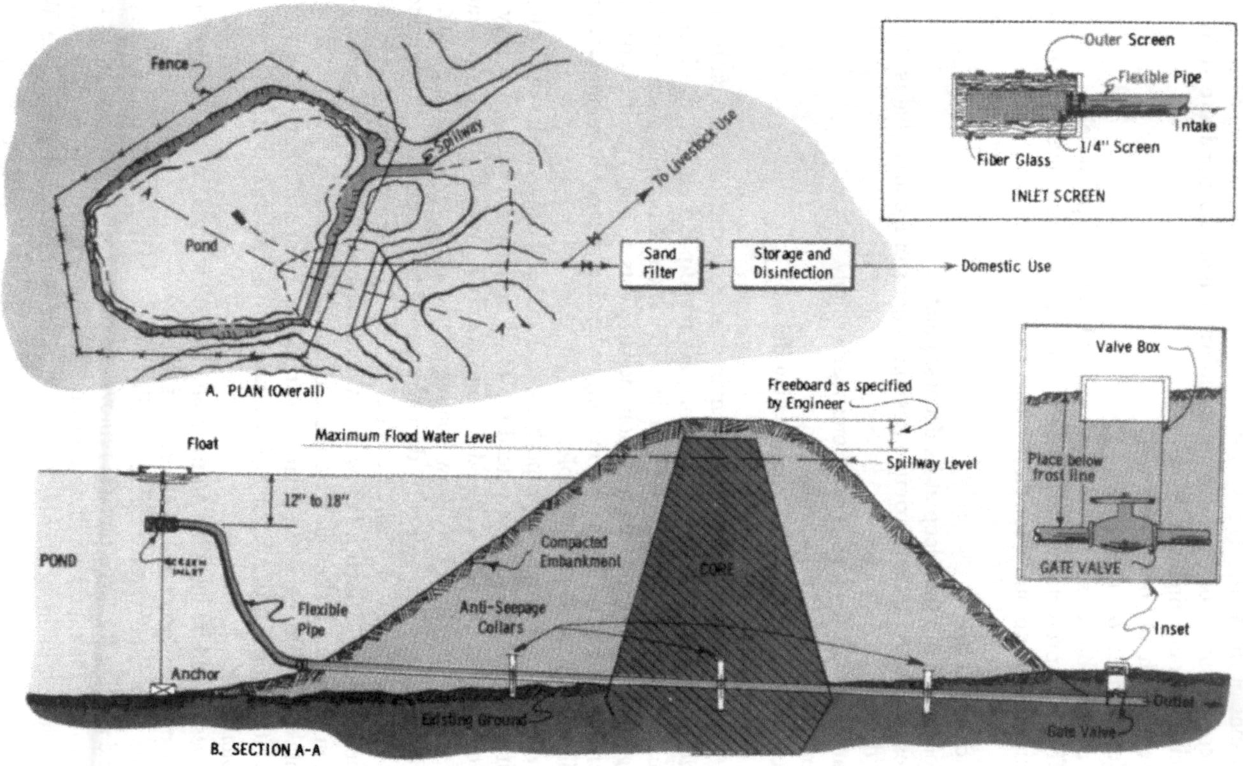

FIGURE 12. - *Pond.*

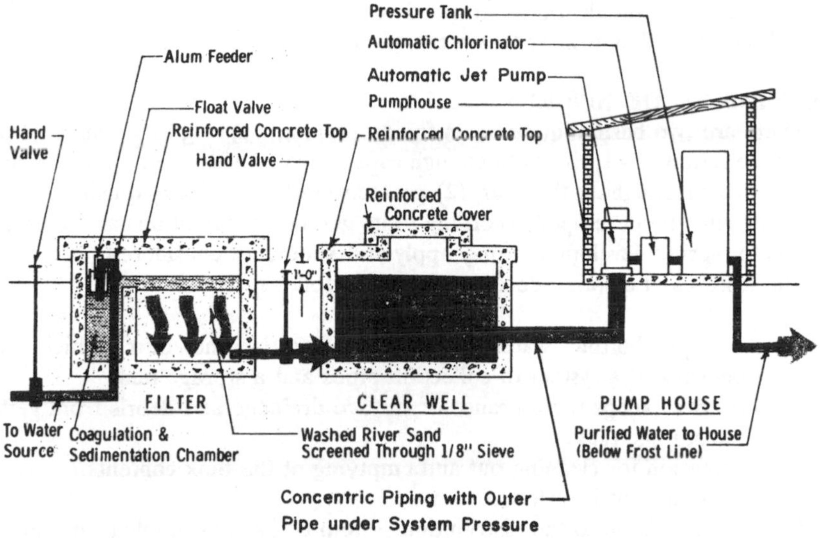

FIGURE 13. - ***Schematic diagram of pond water-treatment system.***

Protection From Freezing. Protection from freezing must be provided unless the plant is not operated or drained during freezing weather. In general, the filter and pump room should be located in a building that can be heated. With the right topography, the need for heating can be eliminated by placing the pump room and filter underground on a hillside. Gravity drainage from the pump room must be possible to prevent flooding. No matter what arrangement is used, the filter and pump room must be easy to get to for maintenance and operation.

Tastes and Odors. Surface water frequently develops musty or unpleasant tastes and odors. These are generally caused by the presence of microscopic plants called algae. There are many kinds of algae. Some occur in long thread-like filaments that are visible as large green masses of scum; others may be separately free floating and entirely invisible to the unaided eye. Some varieties grow in the early spring, others in summer, and still others in the fall. Tastes and odors generally result from the decay of dead algae. This decay occurs naturally as plants pass through their life cycle. For discussion of algae control, see Nuisance Organisms in Part IV.

Streams

In some cases, streams receiving runoff from large, uncontrolled watersheds may be the only source of water supply. The physical and bacteriological quality of surface water varies and may impose very high burdens on the treatment facilities.

Stream intakes should be located upstream of sewer outlets or other sources of contamination. The water should be pumped when the amount of silt in the stream is low. Low water levels in the stream usually means that the temperature of the water is higher than normal and the water is of poor chemical quality. However, maximum silt loads occur during maximum runoff. High-water stages, shortly after storms and the silt has settled, are usually the best times for diverting or pumping water to storage. These conditions vary and should be determined for each individual stream. The following chapter discusses techniques for treatment of surface waters for potable use.

DEVELOPMENT OF SPRINGS

There are two basic requirements for developing a spring as a source of domestic water: (1) selection of a spring with enough capacity to provide the required quantity and quality of water throughout the year, (2) protection of the sanitary quality of the spring. Each spring must be developed according to its geological conditions and sources. State standards for spring development may apply and should be considered.

The features of a spring encasement are as follows:

1. An open-bottom, watertight basin intercepting the source which extends to bedrock or a system of collection pipes and a storage tank.
2. A cover keeps the entrance of surface drainage and debris from getting into the storage tank.
3. Provision for cleaning out and emptying of the tank contents.
4. Provision for overflow.
5. A connection to the distribution system or backup supply (see Figure 14).

Tanks are usually made of reinforced concrete, large enough to capture or intercept as much of the spring as possible. When a spring is located on a hillside, the downhill wall and sides should go down to bedrock, or to a depth that will insure an adequate water level in the tank at all times. "Cutoff walls" of concrete or impermeable clay extending out from the sides of the tank may be used to control the water table around the tank. Construct the bottom part of the tank's uphill wall of stone, brick, or other material that will allow water to move freely into the tank from the aquifer. Backfill of graded gravel and sand will help keep that material in place.

A tank cover should be cast in place to insure a good fit. Forms should be designed to allow for shrinkage of concrete and expansion of lumber. The cover should be extend down over the top edge of the tank at least 2 inches. The tank cover should be heavy enough (so that it cannot be dislodged by children) and should be equipped for locking.

A drain pipe with an exterior valve should be installed close to the wall of the tank near the bottom. The pipe should extend horizontally, clearing the normal ground level at the point of discharge by at least 6 inches. The discharge end of the pipe should be screened to keep out rodents and insects.

The overflow pipe is usually placed slightly below the maximum water-level line and screened. A layer of rock (a "drain apron") should be laid around the point of overflow discharge to prevent erosion.

The spring's outlet should be located about 6 inches above the drain outlet and screened. Care should be taken in fitting pipes through the tank's walls to insure a good bond with the concrete and freedom from "honeycomb" around the pipes.

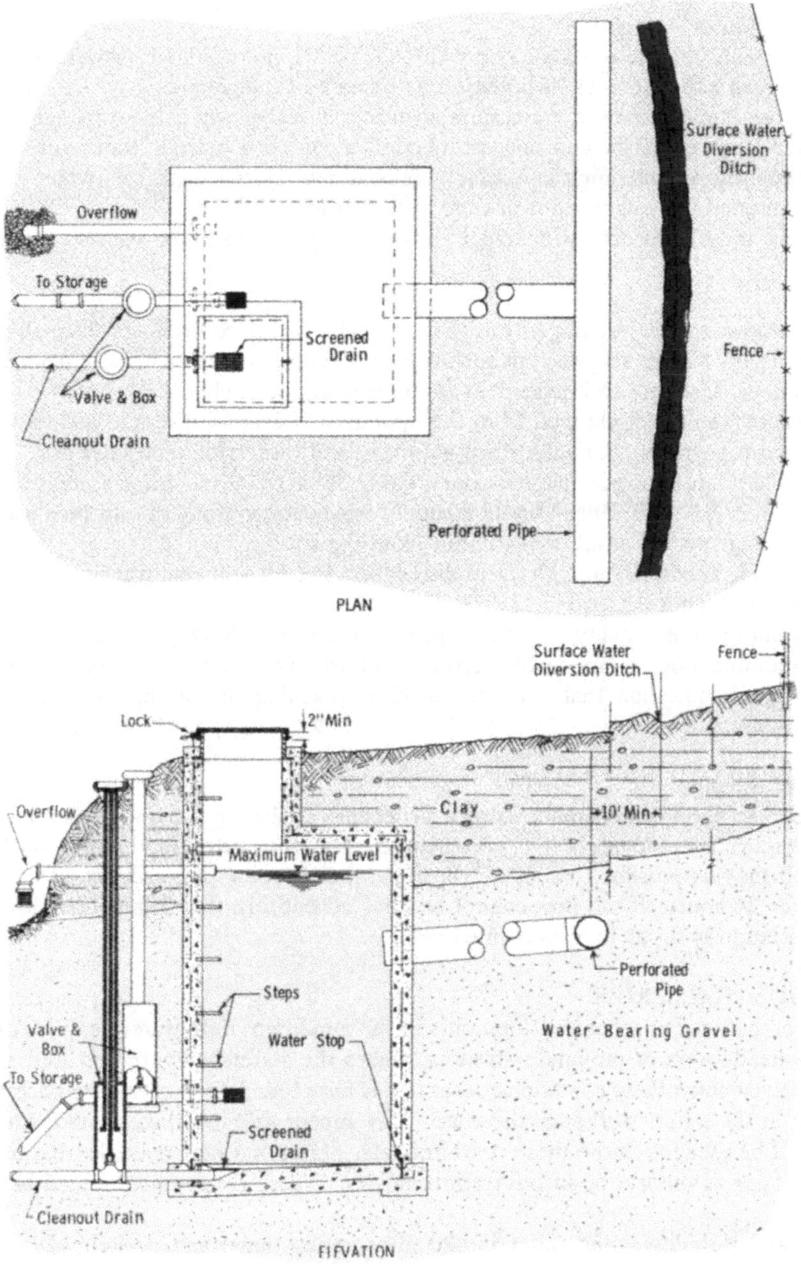
Surface Water Diversion Ditch
Overflow
To Storage
Screened Drain
Fence
Valve & Box
Cleanout Drain
Perforated Pipe
PLAN
Surface Water Diversion Ditch
Fence
Lock
2"Min
Overflow
Clay
10' Min
Maximum Water Level
Perforated Pipe
Steps
Valve & Box
Water Stop
Water-Bearing Gravel
To Storage
Screened Drain
Cleanout Drain
ELEVATION

FIGURE 14. - ***Spring Protection.***

Sanitary Protection of Springs

Springs usually become contaminated when barnyards, pasture lands, sewers, septic tanks, cesspools, or other sources of pollution are located on higher land next to them. In limestone formations, however, contaminated material frequently enters through sink holes or other large openings and may be carried along with ground water for long distances. Similarly, if contamination enters the water in glacial drift, this water may remain contaminated for long periods of time and over long distances.

Following these measures will help to insure a spring water of consistent high quality:

1. Remove surface drainage from the site. Dig a surface drainage ditch uphill from the source, to intercept surface-water runoff and carry it away from the source. Use care and judgement when locating the ditch and the outlet points where water is discharged from the system. Criteria used should include the local topography, the subsurface geology, land ownership, and land use.
2. Build a fence to keep livestock out. Take into account the criteria mentioned in Item 1. The fence should keep livestock away from the surface-water drainage system at all points uphill from the source.
3. Provide access to the tank for maintenance, but prevent removal of the cover by installing locks.
4. Monitor the quality of the spring water regularly by checking for contamination. A noticeable increase in turbidity or flow after a rainstorm is a good indication that surface runoff is reaching the spring and possibly contaminating it.

Disinfection of Springs

Disinfect the spring encasement using a procedure similar to the one used for dug wells. If there is not enough water pressure to raise the water to the top of the encasement, it may be possible to shut off the flow, thus keeping the disinfectant in the encasement for 24 hours. If the flow cannot be shut off entirely, the disinfectant should be added continuously for as long as is practical.

INFILTRATION GALLERIES

Many recreational or other developments in the mountains have water supplies that are near the headwaters of mountain streams, where the watersheds are forested and uninhabited. Even under these conditions, researchers have found disease-causing bacteria and protozoa in the water. Although the water may appear safe to drink, some form of treatment must be used to remove these contaminants. States may have specific treatment requirements such as the use of an infiltration gallery. These requirements need to be considered.

Debris and turbidity at the water intake after spring thaws and periods of heavy rainfall can cause problems in operating and maintaining these supplies. If possible, this material should be removed before it reaches the intake. Experience has shown that debris and turbidity can be removed successfully, especially when small volumes of water are involved, by installing an infiltration gallery at or near the intake.

Where the soil next to a stream will allow water to pass through it, the water can be intercepted by an infiltration gallery located a reasonable distance from the high-water level and a safe distance below the ground surface. Install infiltration galleries so that they will intercept the flow from the stream after it flows through the soil formations.

A typical infiltration gallery installation generally involves the construction of an underdrained, sand-filter trench parallel to the stream bed and about 10 feet away from the high-water mark. The trench should have a minimum width of 30 inches and a depth of about 10 feet. At the bottom of the trench, perforated or open-joint tile should be laid in a bed of gravel about 12 inches thick with about 4 inches of graded gravel over the top to support the sand. Then, the gravel should be covered with clean, coarse sand to a minimum depth of 24 inches, and the rest of the trench backfilled with fairly nonporous material. The collection tile ends in a watertight, concrete basin where it is chlorinated and diverted or pumped to the distribution system.

Where soil formations adjoining a stream will not allow water to flow through them, the debris and turbidity that are occasionally encountered in a mountain stream should be removed using a modified infiltration gallery/slow-sand filter combination in the stream bed. Typical installation involves building a dam across the stream to form a natural pool, or digging a pool behind the dam. The filter should be installed in the pool behind the dam by laying perforated pipe in a bed of graded gravel that is covered by at least 24 inches of clean, coarse sand. About 24 inches of space should be allowed between the surface of the sand and the dam spillway. The collection lines should end in a watertight, concrete basin, next to the upstream side of the dam, where the water is diverted for chlorination.

Where the soil next to a stream will allow water to pass through it, the water can be intercepted by an infiltration gallery located a reasonable distance from the high water level and a safe distance below the ground surface. Install infiltration galleries so that they will intercept the flow from the stream after it flows through the soil formations.

A typical infiltration gallery installation generally involves the construction of a [illegible] sand and gravel filter trench parallel to the stream bed and about 10 feet away from the high water mark. The trench should have a minimum width of 30 inches and a depth of about 10 feet. At the bottom of the trench, perforated or open joint tile should be laid in a bed of [illegible] gravel. The tile should be covered with about 6 inches of graded gravel over the top of the tile and [illegible]. Then the trench should be backfilled with sand [illegible] to a depth of [illegible] and the rest of the trench backfilled with [illegible]. The collection tile [illegible] a watertight [illegible] be used in a [illegible] similar to the [illegible].

Where soil formations adjoining a stream will not allow water to pass through it, the debris and turbidity that are occasionally encountered in a mountain stream should be removed using a modified infiltration gallery, dam, and filter combination in the stream bed. Typical installations involve building a dam across the stream to form a natural pool or digging a pool behind the dam. The filter should be installed in the pool behind the dam by using perforated pipe in a bed of graded gravel that is covered by at least 24 inches of clean, coarse sand. About 24 inches of space should be allowed between the surface of the sand and the dam spillway. The collection lines should end in a watertight concrete basin, next to the upstream side of the dam, where the water is diverted for chlorination.

Part IV

Water Treatment

NEED AND PURPOSE

Raw waters from natural sources may require treatment prior to use. Water supplies may contain pathogenic (disease-producing) organisms, suspended particles, or dissolved chemical substances. Except in limestone areas, ground water is less likely to have pathogenic organisms than surface water, but may contain unpleasant tastes and odors or mineral impurities. Some of these objectionable characteristics may be tolerated temporarily, but the quality of the water should be raised to the highest possible level by treatment. Even when nearly-ideal water can be found elsewhere, it is still a good idea to provide for treatment of the less-desirable source in order to have a backup supply of safe water.

The quality of water constantly changes. Natural processes which affect water quality are the dissolving of minerals, sedimentation, filtration, aeration, sunlight, and biochemical decomposition. Natural processes may either contaminate or purify water, but the natural processes of purification are neither consistent nor reliable.

Bacteria are numerous in waters at or near the earth's surface. Their numbers may be reduced by seeping into the ground, lack of oxygen, or from being underground for long periods under unfavorable conditions for bacterial growth. When water flows through underground fissures or channels, however, it may remain contaminated over long distances and time periods.

The belief that flowing water purifies itself after traveling various distances has led to feelings of false security about its safety. Under certain conditions, the number of microorganisms in flowing water may actually rise instead of fall.

Water treatment incorporates, changes, or adds to certain natural purification processes. Water treatment may condition or reduce to acceptable levels chemicals and impurities which may be present in the water.

Treatment may take place at a treatment facility with treated water pumped to the distribution system or to storage, or may consist of household point-of-use (POU) or point-of-entry (POE) treatment units.

WATER TREATMENT

Some of the natural treatment processes and manmade changes to those processes are discussed in the following section.

Coagulation-Flocculation

Coagulation is the process of forming particles in a liquid by the addition of a chemical such as alum to the water. Iron salts, such as ferric chloride, and organic coagulant polymers can also be used. The chemical is mixed with the cloudy, turbid water and then gently stirred by mechanical or hydraulic means to allow flocculation to take place. During this flocculation stage, the suspended particles will combine physically and form a mass of particles. The larger particles will be able to settle out by gravity. This settling may be done in a separate (sedimentation) tank or in the same tank after the mixing period. Adjustment of pH may be needed after this process, because some coagulant chemicals lower pH. Some colors can also be removed from water using coagulation. Competent engineering advice, however, should be obtained on specific coagulation problems.

Sedimentation

Sedimentation is a process of heavy suspended material in water settling out and collecting on the bottom. Sedimentation usually follows coagulation-flocculation.

This settling action can be done in a still pond or properly constructed tank or basin. It takes a few hours to get a significant reduction in suspended matter. The inlet of the tank should be set up so that the incoming water is spread evenly across the entire width of the tank as the water flows to the outlet at the opposite end. Baffles are usually constructed to even out the flow of water through the tank. Cleaning and repairing the installation can be made easier if the tank is designed with two separate sections, so that one can be used alone while the other is cleaned and/or repaired.

When a water source contains a large amount of turbidity, much of it can be removed by sedimentation. A protected pond with gentle grassy slopes often results in reasonably clear raw water.

Filtration

Filtration is the process of removing suspended matter from water as it passes through beds of porous material. The degree of removal depends on the type and size of the filter media, the thickness of the media bed, and the size and quantity of the suspended matter. Since bacteria can travel long distances through filters, filters alone should not be relied upon to produce bacteriologically safe water. Filtered water must also be disinfected. Types of filtration that may be used include:

Slow Sand Filters. Water passes slowly through beds of fine sand at rates averaging 0.05 gallon per minute per square foot of filter area. Properly constructed slow sand filters do not require much maintenance and can be easily adapted to individual water systems (See Figure 15). The frequency of cleaning will vary depending on the turbidity of the water and other factors. It is necessary to clean the filter regularly by removing about one inch of sand from the surface of the filter and either disposing of it or saving it for washing and reuse. This removal will make it necessary to add new or washed sand.

Sand for slow sand filters should consist of hard, durable grains free from clay, loam, dirt, and organic matter. It should have a "sieve analysis" that falls within the range of values shown in Table 8, adapted from the AWWA manual *Water Treatment Plant Design*.

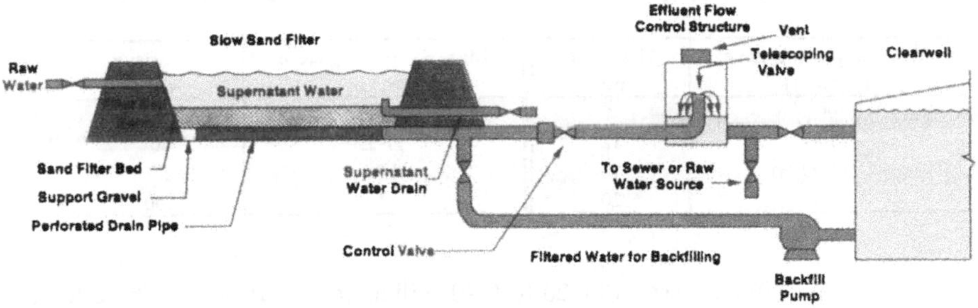

Typical unhoused slow sand filter installation

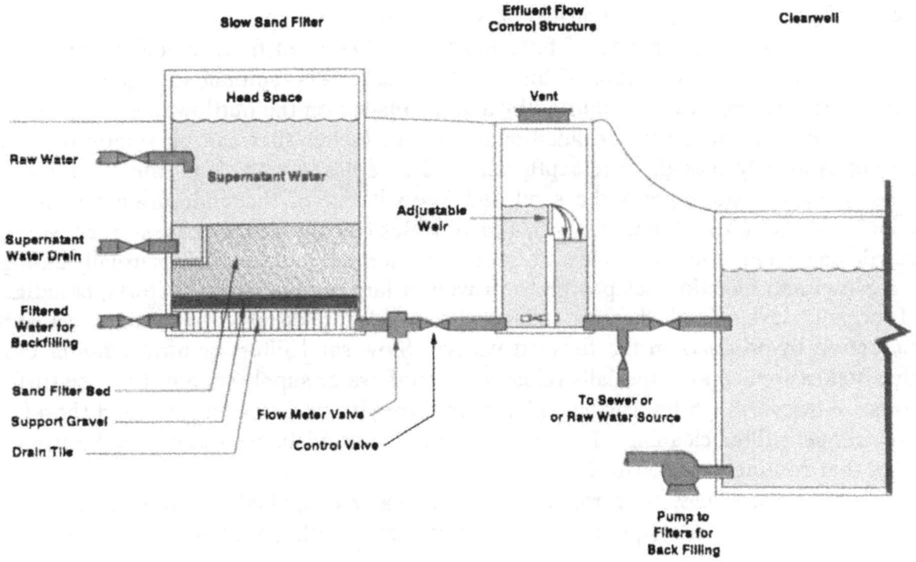

Typical housed slow sand filter installation

FIGURE 15. - ***Slow sand filtration diagram.***

From "Technologiew for Upgrading Existing or Designing New Drinking Water Facilities", EPA (1990)

TABLE 8. - ***Recommended mechanical analysis of slow sand filter media***

Material passing sieve (percent)	U.S. sieve no.	Material passing sieve (percent)	U.S. sieve no.
99	4	33-55	30
90-97	12	17-35	40
75-90	16	4-10	60
60-80	20	1	100

Sands with an "effective size" of 0.20 to 0.40 millimeters work well. The effective size is determined from the dimensions of that sieve or mesh which allows 10 percent of the sample to pass through and retains the remaining 90 percent. This means 10 percent of the grains ar smaller and 90 percent larger. The "uniformity coefficient" should be between 2.0 and 3.0. The coefficient is the ratio of the diameter of a sand grain that is too large to pass through a sieve that allows 60 percent of the material (by weight) to pass through, to the diameter of a grain that is too large to pass through a sieve that allows 10 percent of the material (by weight) to pass through.

For best results, the rate of filtration for a slow sand filter should be 60 to 180 gallons per day per square foot of filter bed surface. The amount of water that flows through the filter bed can be adjusted by a valve placed on the outflow line. Between 27 and 36 inches of sand, with an additional 6 to 12 inches that can be removed during cleaning, is usually enough. The depth removed is replaced with clean sand. Six to eight inches of gravel will support the sand and keep it out of the underdrain system. A 1¼-inch plastic pipe drilled with 3/4-inch holes facing down makes a convenient underdrain system. One to two feet of freeboard on the top of the filter is usually enough.

Slow sand filtration has proven to provide a large reduction in bacteria, cloudiness and organic levels, and thereby reduce the need for disinfection and presence of disinfection by-products in the finished water. Slow sand filters remove *Giardia* cysts, which makes application especially valuable to small water supply systems that use surface waters. However, high levels of cloudiness require slower filtration rates and therefore, more frequent filter cleaning. For that reason, slow sand filters should not be used on waters that routinely exceed 10 NTU.

Rapid Sand Filters. In a rapid sand filter, water is applied at a rate at or above 2 gallons per minute per square foot of filter area, with an allowance for frequent backwashing.

Rapid or "high rate" filters usually follow coagulation and sedimentation for very cloudy source water. Rapid sand filtration is not usually good for very small water supplies because of the necessary controls and additional attention it requires. However, package plants have the necessary filtration processes and automatic controls to meet small system needs.

Rapid sand filters may use a top layer of quartz sand or anthracite coal. Ordinary bank sand cannot be used without screening, washing to remove organic matter and grading to produce the correct size. In addition, it should be free from carbonate hardness.

If the sand is too fine, it will not allow water to pass through it freely, and it will require frequent cleaning. If it is too coarse, it will not effectively remove suspended matter (turbidity).

The sand usually used for rapid sand filters has an effective size of 0.4 to 0.5 mm and a uniformity coefficient between 1.3 and 1.75. Anthracite coal, crushed to slightly larger sizes, is also used instead of sand or to supplement it as a surface layer of lower specific gravity.

Rapid sand filters do not remove bacteria and viruses as well as slow sand filters unless chemical pretreatment (coagulation, flocculation, sedimentation) is optimized. Further treatment, such as disinfection, is required for filters treating surface water.

Pressure Sand Filters. In pressure sand filters, water is filtered at 2 gallons per minute or higher per square foot of filter area. Chemical pretreatment is required and equipment must be provided to allow frequent backwashing of the filter.

Diatomaceous Earth Filters. In diatomaceous earth filters, suspended solids are removed by passing the water through a diatomaceous filter supported by a rigid base support at about the same rates as pressure sand filters.

Diatomaceous earth filters, which require regular attention, are of two types--vacuum and pressure. The filter has several elements, usually small tubes or hollow plates, which are coated with a layer of diatomaceous earth. The water passes through this earth layer, into the tube or plate, and out. A mixture of earth is fed during the filtering run, and when the filter is clogged up enough with suspended matter to slow down the filtering rate, the coating is removed with a backwash. When properly operated and maintained, these filters are effective at removing bacteria and cysts.

Cartridge Filters. Cartridge filters with porous ceramic or glass fiber filter elements, with pore sizes as small as 0.2 μm, may be able to produce drinkable water from raw surface water supplies containing low levels of cloudiness, algae, protozoa and bacteria. Cartridge filters must be used in combination with disinfection. The advantage of a small system is (with the exception of chlorination) that no other chemicals are required. The process is one of physical removal of small particles by straining as water passes through the porous filter. Other than occasional cleaning or filter replacement, they are simple to operate and do not require any special skills or knowledge.

Proper disinfection of water before drinking or cooking is necessary to assure its safety (see "Disinfection" section which follows).

DISINFECTION

Water systems using surface water or water under the direct influence of surface water must disinfect their water supply. In addition, a filtering system is also required, unless otherwise approved by the state agency responsible for water supply since disinfection alone may not protect against some contaminants (such as *Giardia lamblia*). These water treatment guidelines are part of the surface water treatment requirements enacted by EPA to protect drinking water from bacteriological contamination. Disinfection and filtration are both necessary to destroy all harmful bacteria and other organisms. To assure that all harmful organisms are removed, the water to which the disinfectant is added must be low in turbidity. Water with high levels of turbidity may shield microorganisms from the action of disinfectants. After disinfection, water must be stored in suitable tanks or other facilities designed to prevent recontamination.

Emergency disinfection procedures are included in Appendix E.

Chemical Disinfection

The most important features of a chemical disinfectant are: powerful, stable, soluble, non-toxic to man or animals, economic, dependable, lacking residual effects, easy and safe to use and measure, and readily available.

Disinfection must be carefully controlled. Enough disinfectant should be added to water to destroy microorganisms. Too much disinfectant, however, can combine with other compound in water to form harmful chemical disinfection by-products.

Chlorine compounds have some of the best properties of a chemical disinfectant. As a result, chlorine is the most commonly used water disinfectant.

Disinfectant Terminology: Glossary

Chlorine concentration. This is expressed in milligrams per liter (mg/L). One mg/L is equivalent to 1 milligram of chlorine in 1 liter of water. For water, the units of parts per million (ppm) and mg/L are basically the same.

Chlorine feed or dosage. The actual amount in mg/L fed into the water system by the feeder or automatic dosing device is the chlorine feed or dosage.

Chlorine demand. The chlorine fed into the water that combines with impurities, and, therefore, is no longer available for disinfectant action, is commonly called the chlorine demand of the water. Examples of impurities that cause higher chlorine demand are organic materials and certain "reducing" materials such as hydrogen sulfide, ferrous iron, nitrites, etc.

Free and combined chlorine. Chlorine can also combine with ammonia nitrogen, if any is present in the water, to form chlorine compounds that have some ability to disinfect. These chlorine compounds are called combined chlorine residual. However, if no ammonia is present in the water, the chlorine that remains in the water once the chlorine demand has been satisfied is called "free chlorine residual".

Chlorine contact time. The chlorine contact time is the period of time between the addition of chlorine and the use of the water. Chlorine needs a certain amount of contact time in order for it to act as a disinfectant.

Contact time ("T") and chlorine residual concentration ("C") affect how well disinfection will work. For example, surface waters that are not filtered (i.e., only disinfected) require long contact times to assure inactivation of *Giardia* cysts: after 120 minutes of contact, the chlorine residual in a tank should be at least one mg/L. However, the "C" and "T" needed for treating ground water and filtered surface water are different than for unfiltered surface water, and depend on water temperature, pH and other conditions.

Chlorine Disinfection

In general, the primary factors that determine the disinfectant efficiency of chlorine are as follows:

Chlorine concentration. Higher concentrations of chlorine results in effective and faster disinfection. However, there is a point of limited return as higher and higher dosages are used. For that reason, chlorine concentrations should not be more than 5 mg/L in drinking water.

Type of chlorine residual. Free chlorine is a much better disinfectant than combined chlorine.

Temperature of the water in which contact is made. The higher the temperature, the better the disinfection.

The pH of the water in which contact is made. The lower the pH, the more effective the disinfection.

Contact time between the water and chlorine. To ensure proper disinfection, the disinfectant must be in contact with the target microorganisms for a sufficient amount of time. Contact time (CT) values describe the extent of disinfection as a product of the disinfectant residual concentration, C, (in mg/L) and the contact time T, (in minutes). EPA's *Guidance Manual for Compliance With the Filtration and Disinfection Requirements for Public Water Systems Using Surface Water Sources* provides CT values for achieving various levels of *Giardia* and virus inactivation. Sufficient CT is a state regulated requirement of the Surface Water Treatment Rule.

Generally speaking, the longer the contact time, the better the disinfection. As a basic guideline, chlorine residuals and time can be changed to provide acceptable disinfection for water at 10°C and 7.0 pH, if the following CT values are provided:

C x T > 120 for an Unfiltered Surface Water
C x T > 20 for a Filtered Surface Water
C x T > 6 for Groundwater

Where C is the chlorine residual in mg/L, and
T is the contact time in minutes. Either variable, residual or time, can be manipulated to obtain the above products.

Note: These CT recommendations are provided here as a general guideline only. They may not be sufficient for your water. The guidance manual referred to above should be consulted for the exact CT values required.

Chlorine Compounds and Solutions. Compounds of chlorine, such as sodium or calcium hypochlorite, have excellent disinfecting properties. In small water systems, these chlorine compounds are usually added to the water after being dissolved in a solution form, but may also be used in tablet, powder or pellet form.

One of the most commonly used forms of chlorine is calcium hypochlorite. It is available in soluble powder and tablets. These compounds are "high-test" hypochlorite and contain 65 to 75 percent available chlorine by weight. Packed in cans or drums, these compounds are stable and will not break down if properly stored and handled.

Prepared sodium hypochlorite solution is available commercially with a strength of approximately 5 percent available chlorine by weight. Other sodium hypochlorite solutions vary in strength from 3 to 15 percent available chlorine by weight, and are reasonably stable when stored in a cool, dark place. These solutions are diluted with drinkable water to obtain the desired solution strength to be fed into the system.

When hypochlorite powders are used, fresh solutions should be prepared frequently because chlorine solutions will get weaker over time. The container or vessel used for preparation, storage, and distribution of the chlorine solution should be resistant to corrosion and shield the solution from light. (Light causes a reaction that weakens chlorine solutions.) Good materials include glass, plastic, crockery, or rubber-lined metal containers.

Hypochlorite solutions can be used either at full strength or diluted to a strength suited to the feeding equipment and the rate of water flow. In preparing these solutions, one must take into account the chlorine content of the concentrated (stock) solution. For example, if 5 gallons of 2 percent solution are to be prepared with a high-test calcium hypochlorite powder or tablet containing 70 percent available chlorine, 1.2 pounds of high-test hypochlorite would be needed.

Pounds of compound required:

$$= \frac{\text{\% strength of solution} \times \text{gallons of solution needed} \times \text{8.3 lbs./gal of water}}{\text{\% available chlorine in compound}}$$

$$= \frac{2 \times 5 \times 8.3}{70}$$

$$= 1.2 \text{ pounds}$$

Similarly, the following formula can be used to determine the pounds of chlorine required per day to disinfect at a particular dose:

Chlorine, lbs/day = (Dosage, mg/L)(8.3 lbs/gal of water)(Flow or quantity, mgd[1])

For example, how many pounds of chlorine would be required to feed a dose of 2.0 mg/L to 50,000 gallons of water?

Chlorine, lbs/day = (2.0 mg/L)(8.3 lbs/gal)(.05 mgd)

Chlorine, lbs/day = 2

Now, if we used calcium hypochlorite that was 65 percent available chlorine, how many pounds of this compound would be required to satisfy the previous condition?

$$\text{Calcium hypochlorite, lbs/day} = \frac{\text{Chlorine, lbs/day}}{\text{Percent Available}}$$

$$\text{Calcium hypochlorite, lbs/day} = \frac{2}{65\%} \text{ or } \frac{2}{0.65}$$

Calcium hypochlorite, lbs/day = 3 pounds

---

1 mgd = million gallons per day

For a hypochlorite solution, how many gallons of 2 percent sodium hypochlorite solution would be required to feed a 2.0 mg/L dosage to our 50,000 gallons of water?

$$\text{Hypochlorite solution, (gal/day)} = \frac{(\text{Chlorine, lbs/day})(100\%)}{(\text{Hypochlorite strength, }\%)(8.3 \text{ lbs/gal})}$$

$$\text{Hypochlorite solution, (gal/day)} = \frac{(2 \text{ lbs/day})(100\%)}{(2\%)(8.3 \text{ lbs/gal})}$$

$$\text{Hypochlorite solution, (gal/day)} = \frac{200}{16.6} = 12$$

Determination of Chlorine Residual. Residual chlorine can exist in water as a free or combined chlorine residual. When present as a combined chlorine residual, it is combined with either organic material or ammonia. The sum of combined and free chlorine is called total chlorine residual. Sufficient chlorine is the amount needed to produce a desired residual after a definite contact period whether the chlorine residual is combined, free, or total.

The amount of remaining chlorine (chlorine residual) in the water is normally measured using a simple test called the DPD calorimetric test (short for the chemical name N,N-dimethyl-p-phenylene-diaminc). The test can be done outside of a laboratory using special pills that are placed in a test tube. The free or total chlorine residuals produce a violet color that can be compared with a color chart to determine the quantity of chlorine in the water. The kits come with all necessary test tubes, chemicals, color chart and instructions. Firms that specialize in the manufacture of water testing equipment and materials can supply them.

A combination DPD and pH kit is also available at a modest price. State and county water supply agencies can provide names of water test kit suppliers.

Wherever chlorination is needed for disinfection, testing for chlorine residual should be done at least daily, and a daily record of results kept.

For more information on the DPD test and other water testing procedures, descriptions are included in *Standard Methods for the Examination of Water and Wastewater.*[2]

Chlorination Equipment

Hypochlorinators. Hypochlorinators pump or inject a chlorine solution into the water. When they are properly maintained, hypochlorinators are a reliable way to apply chlorine to disinfect water. However, protection against over-feeding or siphoning should be included.

Types of hypochlorinators include positive displacement feeders, aspirator feeders, suction feeders, tablet hypochlorinators and dry pellet chlorinators.

Positive Displacement Feeders. A common type of positive displacement hypochlorinator is one that uses a piston or diaphragm pump to inject the solution. It is adjustable during operation, and can be designed to give reliable and accurate feed rates. When

---

2 Obtainable from the American Waterworks Association, 6666 West Quincy Avenue, Denver, Colorado 80235, Phone: (303) 794-7711.

electricity is available, the stopping and starting of the hypochlorinator can be timed with the water supply pumping unit. This kind of hypochlorinator can be used with any water system, but it is especially useful in systems where water pressure is low and changeable.

Aspirator Feeders. The aspirator feeder uses a vacuum created when water flows either through a venturi tube or perpendicular to a nozzle, to draw chlorine solution from a container into the chlorinator unit, where it is mixed with water. The mixed solution is then injected into the water system. In most cases, the water inlet line to the chlorinator is connected to the discharge side of the water pump, with the chlorine solution being pumped back into the suction side of the same pump. The chlorinator operates only when the pump is operating. Chlorine solution flow rate is controlled by a valve, but pressure changes may cause changes in the feed rate.

Suction Feeders. One type of suction feeder consists of a single line that runs from the chlorine solution container through the chlorinator unit, and connects to the suction side of the water pump. The chlorine solution is pulled from the container by suction created by the water pump.

Another type of suction feeder works on the siphon principle, with the chlorine solution being added directly into the well. This type also consists of a single line, but the line ends in the well, below the water surface, instead of at the suction side of the water pump. When the pump is running, the chlorinator is turned on, so that a valve is opened and the chlorine solution flows into the well.

In each of these units, the chlorine solution flow rate is regulated by a control valve, and the chlorinator runs only when the pump is running. The pump circuit should be connected to a liquid level control so that the water supply pump shuts off when the chlorine solution is gone.

Tablet Hypochlorinators. The tablet ("erosion type") hypochlorinating unit consists of a special pot feeder containing calcium hypochlorite tablets. Accurately controlled by an inlet baffle or flowmeter, small jets of feed water flow through the lower portion of the tablet bed. The tablets slowly dissolve, providing a continuous source of fresh hypochlorite solution. Since this unit produces fairly high dose rates, a by-pass may be needed so that only part of the total flow is treated. This type of chlorinator is used when electricity is not available, but it needs good maintenance for efficient operation. It can also operate where the water pressure is low.

Drop Pellet Chlorinators. Drop-feed pellet chlorinators consist of a motorized feeder that drops a controlled number of dry calcium hypochlorite pellets into well water over a period of time. The device can be timed to drop pellets more or less often, depending on the volume of water being treated. This unit runs only when the well pump is running.

Gaseous Feed Chlorinators. In situations in which large quantities of water are treated, chlorine gas in pressure cylinders may be used as the disinfectant. Its use in very small water supply systems may be limited because of its higher cost and the greater safety precautions necessary to guard against accidents. Gaseous chlorine is an extremely hazardous substance.

Solution Supply Monitor. Sensing units can be placed in solution containers to sound a warning alarm when the solution goes below a predetermined level. This equipment can also be connected to the pump so that when the chlorine is about to run out, the pump will automatically shut and activate a warning bell. On that signal, the operator will have to refill the solution container and take necessary steps to ensure proper disinfection.

Chlorine Contact Tanks. Enough time for the chlorine to properly disinfect water for drinking can be provided using a chlorine contact tank. Although mixing is not required, special measures should be taken to ensure that, at maximum flow, 90 percent of the water discharged from the contact tank has been in the tank for the required time period. More information on disinfectant dosages and required contact time can be found in the previous section on "Chlorine Disinfection".

It is important to note that contact time is not simply the required time multiplied by the flow rate. Short-circuiting may occur, in which some water flows straight through the tank, while other water becomes trapped in stagnant zones, remaining in the tank for a long time. This can be corrected by using a properly designed tank, possibly with baffles to direct the water flow.

Chlorination Control

Several factors have a direct bearing on the effectiveness of chlorine. Because of these factors, it is not possible to suggest rigid standards that will work in all water supply systems. It is possible, however, to offer some general guidelines for water supply operation and maintenance.

Simple Chlorination. Unless bacteriological or other tests indicate the need to maintain higher minimum concentrations of free residual chlorine, at least 0.2 mg/L of free residual chlorine should be in contact with treated ground water (at 10°C) for at least 30 minutes before the water reaches the first user from the point of where the chlorine is added. As noted above, surface water supplies, and different water temperatures, would require different disinfectant concentrations and contact time. It is a good idea to maintain a detectable free chlorine residual at the end of the distribution system when using simple chlorination. Refer to the section on "Chlorine Disinfection" for more information on establishing the right chlorine dosages.

A method known as superchlorination-dechlorination can be used to solve the problem of insufficient contact time in a water system. By this method, chlorine is added to the water in increased amounts (superchlorination) to provide a minimum chlorine residual of 3.0 mg/L for a minimum contact period of 5 minutes. Removal of the excess chlorine (dechlorination) is then used to get rid of unpleasant chlorine taste.

Records. Proper control of water quality also depends on keeping accurate operating records of chlorination. Those records can indicate chlorination is being done properly, and as a guide for improving operations. The record should show the amount of water treated, amount of chlorine used, setting of the chlorination equipment, time and location of chlorine tests, and results of chlorine tests. This information should be kept current and posted near the chlorination equipment.

DISINFECTION WITH ULTRAVIOLET LIGHT

Ultraviolet (UV) light produced from UV lamps has been shown to be an effective disinfectant. In disinfecting water, the amount of UV radiation needed depends on factors such as turbidity, color, and dissolved iron salts, which prevent ultraviolet energy from entering the water. For that reason, UV light would not work for disinfecting turbid water and is normally used only in treating groundwater.

Cylindrical UV units with standard plumbing fittings have been designed for use in water lines. They should be checked often for light intensity and cleaned of any material that would block radiation from reaching the water. An advantage of disinfection with UV light is that the equipment is readily available and easy to operate and maintain. A disadvantage of UV light is that it does not provide a residual disinfectant in the water (as does chlorine). Thus, there is no protection against recontamination in UV-disinfected water and another disinfectant may be needed (such as chlorine) to maintain a bacteria controlling residual in the distribution system. Another disadvantage of UV is that it is not effective against some microorganisms, such as *Giardia lamblia* cysts. Where *Giardia lamblia* are a potential problem, such as with surface water sources, disinfection by UV is not recommended.

DISINFECTION WITH OZONE

Ozone is the strongest oxidizing agent available for water treatment. It is a very strong disinfectant, and can oxidize the following types of compounds: taste and odor compounds, certain organics, iron and manganese, and sulfides. However, like UV disinfection, ozone does not provide a residual to protect against recontamination.

Either air or oxygen may be used to generate ozone. A typical ozone system using air as the feed source would consist of the following units: air filter, air compressor, air cooling and drying system, the ozone generator where ozone is created from oxygen in the air as it passes through a high voltage electric current, a contact chamber where the ozone mixes through the water, and a destruction device for the off-gases from the contactor. If oxygen from tanks is used, the air compressor and drying system are not needed.

The contactor for a small water treatment plant is usually a tall chamber or tall fiberglass columns. Contactors are normally designed for a contact time of ten minutes.

Ozone gas leaks in and around a generator and contactor are very dangerous. Ozone is a toxic gas and requires operator protection similar to that for gas feed chlorination systems.

MEMBRANE TECHNOLOGIES

Reverse Osmosis (R.O.) and electrodialysis reversal (EDR) are two commonly used membrane processes for the physical separation and removal of unwanted water contaminants. Membranes act as molecular filters to remove most dissolved minerals, biological organisms and suspended matter from water. Before installing an R.O. or EDR system, care must be taken so that the liquid waste streams can be disposed of properly and that state and local permits are obtained.

Reverse Osmosis

Reverse osmosis (R.O.) is a membrane process that is sometimes referred to as hyperfiltration, the highest form of filtration possible. It uses a special, semi-permeable membrane which, under pressure, permits pure water to pass through it while acting as a barrier to dissolved salts and other impurities.

A basic R.O. system consists of a high pressure pump, control system to regulate flows and pressure, and the R.O. module. Normally, some form of pretreatment is used. This can be anything from a simple filter to the combined use of coagulation, sedimentation and filtration followed by pH adjustment when treating very dirty water. Post-treatment methods can include pH adjustment, aeration and chlorination for disinfection.

Electrodialysis Reversal

The electrodialysis reversal (EDR) process is based on another membrane technology, electrodialysis. Using alternately placed positively and negatively charged membranes allows the passage of ions. Separate areas of ion-rich and ion-poor water are created. The membranes are arranged in stacks, and the desired salt or other contaminant reduction is achieved by passing the water through the proper number of stacks. With EDR, the DC current is reversed periodically, typically two to four times per hour. This causes ion movement to reverse direction, and flushes the scale-forming ions from membrane surfaces. This lengthens membrane life and reduces maintenance and replacement costs.

AERATION

Aeration is the process of bringing air into contact with a liquid such as water. Many methods can be used for effective aeration, including spraying water into the air, allowing water to fall over a spillway in a turbulent stream, or letting water trickle in multiple streams or droplets through a series of perforated plates or through a packed tower. Although aeration may be performed in an open system, care should be taken to prevent possible external contamination of the water. Whenever possible, a totally enclosed system should be used.

Aeration may be used to oxidize iron or manganese and remove odors from water, such as those caused by hydrogen sulfide and algae. Aeration also efficiently removes radon gas and volatile organic compounds (VOC's) from water. It is also effective in increasing the oxygen content of water that is too low in dissolved oxygen. The flat taste of cistern water and distilled water may be improved by adding oxygen. Carbon dioxide and other gases, that increase the corrosiveness of water, can be largely removed by aeration. However, aeration increases the level of oxygen, increasing corrosion, which may partially offset the advantage of decreasing the carbon dioxide levels.

Aeration partially oxidizes dissolved iron and manganese, changing iron into its insoluble form. Sometimes a short period of storage permits the insoluble particles to settle out; at other times, the particles of iron and manganese must be filtered out.

A simple cascade device or a coke tray (wire-bottom trays filled with activated carbon) aerator can be added to a water supply system. In addition to aerating, the coke tray will reduce tastes and odors.

Insects such as the chironomus fly may lay eggs in the stagnant portion of the aerator tray. Covering the aerator prevents flies from getting into the aerator. Screening will also provide protection from windblown debris.

OTHER TREATMENT

Iron and/or Manganese Removal

Iron and/or manganese in water creates a problem common to many individual water supply systems. When both are present beyond recommended levels, special attention should be paid to the problem. The removal of iron and manganese depends on the type and quantity, and this helps determine the best procedure and (possibly) equipment to use.

Well water is usually clear and colorless when it comes out of the faucet or tap. When water containing colorless, dissolved iron is allowed to stand in a cooking container or comes in contact with a sink or bathtub, the iron combines with oxygen from the air to form reddish-brown particles (commonly called rust). Manganese forms brownish-black particles.

These impurities can give a metallic taste to water or to food. Deposits of iron and manganese produce rusty or brown stains on plumbing fixtures, fabrics, dishes, and utensils. Soaps or detergents will not remove these stains, and bleaches and alkaline builders (often sodium phosphate) can make it worse. After a long period, iron deposits can build up in pressure tanks, water heaters, and pipelines, reducing the quantity and pressure of the water supply.

Iron and manganese can be removed through chlorination and filtration. The chlorine chemically oxidizes the iron or manganese (forming a particle), kills iron bacteria, and kills any disease bacteria that may be present. The filter then removes the iron or manganese particles. Other techniques, such as aeration followed by filtration, ion exchange with greensand, or treatment with potassium permanganate followed by filtration, will also remove these materials.

Insoluble iron or manganese and iron bacteria will clog up the mineral bed and the valving of a water softener. It is best, therefore, to remove iron and manganese before the water reaches the softener.

When a backwash filter medium is used, an adequate quantity of water at a high enough pressure must be provided for removing the iron particles.

Potassium permanganate can be used in place of chlorine. The dose, however, must be carefully controlled. Too little permanganate will not oxidize all the iron and manganese, and too much will allow permanganate to enter the distribution system and cause a pink color.

Iron can also be oxidized, by simple aeration, to removable ferric hydroxide by exposing water to air in sprays or by cascading (water falls) over steps or trays. The ferric hydroxide is then filtered out. The main advantage of this method is that it requires no chemicals.

If the water contains less than 1.0 mg/L iron and less than 0.3 mg/L manganese, using polyphosphates followed by chlorination can be effective and inexpensive. Below these concentrations, the polyphosphates combine with the iron and manganese preventing them from being oxidized. Any of the three polyphosphates (pyrophosphate, tripolyphosphate, or metaphosphate) can be used.

To determine the best polyphosphate to use and the right dosage, prepare a series of samples at different concentrations. Add chlorine, and observe the samples daily against a white background. The right polyphosphate dose is the lowest dose that does not noticeably discolor the water samples for four days.

Iron Bacteria Removal

Under certain conditions, the removal of iron compounds from a water supply may be more difficult due to the presence of iron bacteria. When dissolved iron and oxygen are present in the water, these bacteria get energy from the oxidation of the iron. These bacteria collect within a gelatinous mass which coats underwater surfaces. A slimy, rust-colored mass on the inside surface of flush tanks or water closets is caused by iron bacteria.

Iron bacteria can reduce the flow within water pipes by increasing friction. They may give an unpleasant taste and odor to the water, discolor and spot fabrics and plumbing fixtures, and clog pumps. A slime also builds up on any surface that the water containing these organisms touches. Iron bacteria may be concentrated in a specific area and may periodically break loose and appear at the faucet in visible amounts of rust. Lines may need to be regularly flushed through hydrants or other valves when such red, rusty water is found.

Iron-removal filters or water softeners can remove iron bacteria; however, they often become clogged by the slime. A disinfecting solution should be injected into the water to control the growth of iron bacteria. This causes iron particles to form. These particles can then be removed with a suitable fine filter.

Water Softening

Water softening is a process for removing minerals (primarily calcium and magnesium) that cause hardness. It can also remove small quantities of iron and manganese.

It should be noted that the two softening processes discussed in this section will also remove radium. Where radium is removed, care must be taken to safely dispose the treatment wastes.

Softening of hard water is desirable if:

- Large amounts of soap are needed to produce a lather.
- Hard scale is formed on cooking utensils or laundry basins.
- Hard, chalk-like formations coat the interiors of piping or water tanks.
- Heat-transfer efficiency through the walls of the heating element or exchange unit of the water tank is reduced.

The buildup of scale will reduce the amount of water and pressure a pipe can carry. Excessive scale from hard water is also unpleasant to look at. Experience shows that hardness values much higher than 200 mg/L (12 grains per gallon) may cause some household problems.

Water may be softened by either the ion-exchange or the lime-soda ash process, but both processes increase the sodium content of the water and may make it unfit for people on a low-sodium diet.

Ion Exchange. The ion-exchange process replaces calcium or magnesium ions with sodium ions. The process takes place when hard water containing calcium or magnesium compounds comes in contact with an exchange medium. The materials used in the process of ion exchange are insoluble, granular materials that can perform ion exchange. Ion-exchange material may be classed as follows: glauconite (or greensand); precipitated synthetic, organic (carbonaceous), and synthetic resins, or gel zeolites. The last two are the most commonly used for centralized treatment of drinking water.

The type of ion-exchange material to use depends on the type of water treatment needed. For example, when a sodium zeolite is used to soften water by exchanging sodium ions for calcium and magnesium ions in the hard water, there will eventually be too few sodium zeolite ions to complete the exchange. The hardness of the water entering the unit and the water leaving should be checked regularly. After a certain period of time (determined by the exchange rate), the sodium zeolite must be recharged. The sodium ions are restored to the zeolite by passing a salt (NaCl) or brine solution through the ion-exchange media (bed). The salt solution used must contain the same type of ions (sodium) that were replaced by the calcium and magnesium. The salt solution reverses the ion-exchange process, restoring the sodium zeolite to its original condition.

The type of recharge material or solution that must be used depends upon the type of exchange material in the system.

The ion-exchange method of softening water is fairly simple and can be used easily by a small or individual water supply system. Only part of the hard water needs to be passed through the softening process because it can produce water of zero hardness which could be more corrosive to the metals in the piping and household plumbing. The processed water can then be mixed back in with hard water to produce a final water with a hardness of 50 to 80 mg/L (3 to 5 grains per gallon). Water with a turbidity of more than 5 NTUs (a measure of the cloudiness of water) should first be properly treated for removal of particles to increase the effectiveness and the efficiency of the later softening process.

If the bed turns an orange or rusty color, iron fouling is becoming a problem. Increasing the time of the recharge stage may help. A chemical cleanser (sodium bisulfite) can also be used to remove heavy iron coatings from the media. The cleaner can be added to the regenerating material or mixed in solution form and poured into the softener when the unit is out of service. The softener should be rinsed before using it again after this cleaning.

Ion-exchange softeners are commercially available for individual water systems. Their capacities range from about 85,000 to 550,000 milligrams of hardness removed for each cubic foot of ion-exchange material. Water softeners should only be installed by qualified persons following the manufacturer's instructions and applicable codes. The materials and workmanship should be guaranteed for a specified period of time. First consideration should be given to companies that provide responsible servicing dealers permanently located as close as possible to the water supply system. *Note:* Zeolite softening is not recommended if any of the water consumers are on a restricted sodium diet for medical reasons.

Lime-Soda Ash Process. The use of the lime-soda ash process or the addition of other chemicals may not be practical for a small water supply system. However, they may be able to obtain package plants that soften water. Water used for laundry may be softened at the time of use by adding certain chemicals such as borax, washing soda, trisodiumphosphate, or ammonia. Commercial softening or water conditioning compounds of unknown composition should *not* be used in water intended for drinking or cooking without the advice of the state or local health department.

Fluoride Removal

Although fluoride is an accepted and useful preventive treatment for reducing tooth decay, excessive levels of fluoride may cause pitting and staining of children's teeth or contribute to the development of crippling skeletal fluorosis when consumed over a long period of time. If a water source contains fluoride above 4 mg/L, a fluoride removal treatment should be used.

Two common methods of fluoride removal are reverse osmosis and activated alumina. Reverse osmosis removes the fluoride from the water by using a semi-permeable membrane (see Reverse Osmosis, Page 90). Activated alumina is a commercially available adsorption/ion exchange media which can remove up to 90 percent of fluoride from water. However, it requires an extended contact time and increases the pH to a level that may be unacceptable. Either method may be employed as a reliable means of removing fluoride.

Tastes and Odor Control

Tastes and odors in an individual water supply system fall into two general classes--natural and man-made. Some natural causes may be traced to algae, sulfate-reducing bacteria, leaves, grass, decaying vegetation, dissolved gases, and slime-forming organisms. Some of the man-made causes of taste and odor may be due to chemicals and sewage.

Water having a "rotten egg" odor indicates the presence of hydrogen sulfide and is commonly called sulfur water. In addition to its unpleasant odor, sulfur water may cause a black stain on plumbing fixtures. Hydrogen sulfide is corrosive to common metals and will react with iron, copper, or silver to form the sulfides of these metals.

Depending on the cause, taste and odor can be removed or reduced by aeration or by treatment with activated carbon, copper sulfate, or an oxidizing agent such as chlorine, ozone, or potassium permanganate. It should be noted that chlorine may actually increase, rather than reduce, tastes and odors.

Aeration. Aeration for control of tastes, odors and other substances is described in the section above, entitled "Aeration." Hydrogen sulfide can be removed by a combination oxidization-filtration process. A simple iron-removal filter will also do a good job if there are only small amounts of hydrogen sulfide.

Activated Carbon. Activated carbon treatment involves passing the water to be treated through granular carbon, or adding powdered activated carbon to the water. Activated carbon adsorbs (attracts to itself) large quantities of dissolved gases, soluble organics, and finely divided solids. It controls taste and odor well. Activated carbon can be used in carbon filters available from manufacturers or producers of water-conditioning or treatment equipment. The recommendations included with the filter should be followed.

Copper Sulfate. The most frequent source of taste and odors in an individual water supply system is algae, whose biological byproducts cause taste and odor problems. These tastes and odors may increase when chlorine is added to the water. When algae is present in a water supply, its growth can be controlled by adding copper sulfate to the water source, as described in the section entitled "Nuisance Organisms".

Because algae and other chlorophyll-containing plants need sunlight to live, the storage of water in covered reservoirs inhibits their growth.

Chlorine. Chlorine may be effective in reducing some tastes and odors in water. The process is the same as that described in the section dealing with "Superchlorination-Dechlorination".

Hydrogen sulfide odors in water may be caused by bacterial activity in distribution lines or in water heaters. A maintained chlorine residual in the water distribution system, and periodic disinfection followed by flushing of water heater tanks, may solve the problem.

Corrosion Control

The control of corrosion is important to continuous and efficient operation of the individual water system and to delivery of properly-conditioned water that has not picked up traces of metals, such as lead (from solder and plumbing fixtures), that may be hazardous to health. Whenever corrosion is lessened, maintenance and replacement costs of water pipes, water heaters, or other metal parts of the system are reduced.

Corrosion is commonly defined as an electrochemical reaction in which metal wears away or is destroyed by contact with elements such as air, water, or soil. The important characteristics of water that may affect its corrosiveness to metals include the following:

Acidity. A measure of the water's ability to neutralize alkaline materials. Water with acidity or low alkalinity tends to be corrosive.

Conductivity. A measure of the amount of dissolved mineral salts. An increase in conductivity promotes the flow of electrical current and increases the rate of corrosion.

Oxygen Content. Oxygen dissolved in water promotes corrosion by destroying the thin protective hydrogen film that is present on the surface of metals in water.

Carbon Dioxide. Carbon dioxide forms carbonic acid, which tends to attack metal surfaces.

Water Temperatures. The corrosion rate increases with water temperature.

Corrosion and Scale Relationship

Corrosion and scale often occur together, but they should not be confused. The effect of corrosion is to destroy metal; scale, on the other hand, clogs open sections and line surfaces with deposits. The products of corrosion often add to scale formation, thus making the problem of treating corrosion more difficult.

Prevention of Corrosion

When corrosion is caused by acidity, it can be controlled by an acid neutralizer. Another way to control corrosion is to feed a small amount of commercially available film-forming material into the system. Other methods of controlling corrosion are the installation of dielectric or insulating unions, reduction of water velocities and pressures, removal of oxygen or acid, chemical treatment to decrease acidity, or the use of nonmetallic piping and equipment.

Physical Control. Physical control of corrosion includes proper grounding and system flushing. Pipe surfaces may be attacked by electrolysis, biological growth, or general chemical reactions. Electrolysis is the result of the effects of grounded electrical equipment in the area, and normally corrodes the outside of a water pipe faster than the inside. Electrical equipment should not be grounded to plumbing, since this can greatly increase corrosion. Proper grounding, however, reduces this kind of corrosion. Flushing removes bacteria from the system that can stimulate corrosion. Lead-free or low-lead solder prevents corrosive water from leaching lead into the system.

Chemical Control. The pH of water can be increased with a neutralizing solution, so that it no longer attacks parts of the water system or contributes to electrolytic corrosion. Neutralizing solutions may be prepared by mixing soda ash (58 percent light grade) with water - 3 pounds soda ash to 4 gallons of water. This solution can be fed into the water supply with feeders as described under "Chlorination," and can be mixed with chlorine solutions to both correct pH and disinfect. Soda ash is available at chemical supply houses.

A dosage of lime, soda ash, or caustic soda can be used to adjust the pH of the treated water and deposit a thin film of calcium carbonate on pipe surfaces. This thin coating prevents corrosion of the metal pipe. The right pH for this kind of protection is usually in the range of eight to nine.

The right dosage of film-forming materials is most often determined from past experience with similar water. A dosage of several milligrams per liter is normally required.

Another method of chemical control is the use of a limestone contactor, or neutralizing tank. The limestone contactor looks like a water softener tank, except that it contains limestone or marble chips. Acidic water reacts with these materials and slowly dissolves them until they need to be replaced. This slow dissolving of the limestone neutralizes the water and makes it less corrosive.

The flow rate through a limestone contactor needs to be slow, to allow time for the limestone to neutralize the water. If it becomes necessary to use two contactors, they should be connected in parallel to provide the most neutralizing capacity.

Maintenance of a contactor requires backwashing, sometimes as often as every other day, to loosen and clean the limestone. Because the limestone is heavy, high backwash flows may be necessary. The limestone should be inspected annually and stone added to replace the dissolved stone.

Corrosion control chemicals are the last to be added during water treatment. This is mainly because the pH for successful coagulation and chlorination is much lower than the pH for noncorrosive water. As another example, a calcium carbonate film can form on the filter media and would interfere with proper filtration.

Nuisance Organisms

Organisms that are known to cause problems in water supplies include several species of algae, protozoa, and diatoms that produce tastes and odors and clog filters. Iron bacteria plug water well intakes and clog pipes in distribution systems (see Iron Bacteria Removal, Page 92). Other nuisance organisms are copepoda, whose eggs can pass through filters; midge larvae or bloodworms; and snails and mollusca. These organisms vary in complexity and size. They are uncommon or absent in ground water, but are common in surface water.

Presence of nuisance organisms may not be dangerous to health. Interference with water treatment processes, and unpleasant taste, odor and appearance are the chief complaints against them. The following suggested treatment process can control these organisms, but may also cause harmful conditions if not done properly. Also, before adding chemicals to a surface water, remember to check with local water protection officials and secure any necessary permits.

Control of Algae. Algae can be controlled by treating the water with copper sulfate (blue stone or blue vitriol) or, when possible, by covering the water storage unit to block sunlight. Maintenance of an adequate chlorine residual will control the growth of algae and other organisms as long as the storage unit is covered and protected from contamination. The particular control method, or combination of methods, should be chosen by studying each case to assess the probability of success and the cost involved.

Copper sulfate has been successfully used for the control of algae since 1900. Temperature, pH and alkalinity all affect the solubility of copper in water, so that the dosage needed depends on the chemistry of the water treated and how effective copper is against the particular nuisance organism. Dosage rates of 1 ounce of copper sulfate ($CuSO_4 \cdot 5H_2O$) for each 25,000 gallons of water have worked where the total alkalinity of the water does not exceed 40 mg/L (40 ppm). For more alkaline waters, the dosage can be increased to 5.5 pounds of copper sulfate per acre of surface water treated regardless of depth. Caution should be taken when adding copper sulfate to water since high concentrations (in excess of those suggested to control algae) can kill fish or even be unsafe for human consumption. The copper level should not exceed 1.3 mg/L.

Frequency of treatment depends on temperature, amount of sunlight, and nutrients in the water. Applying the correct amount of chemical regularly over the entire surface area will keep serious algal blooms from coming back. Several treatments per season are generally required, with treatments as frequent as twice a month during the growing season.

The most practical method of application for small ponds is by spraying it on the surface. Or, a burlap bag of copper sulfate can be dragged through the water. Rapid and even spreading of the chemical is important.

It should be noted that the sudden death of heavy growths of algae may be followed by decomposition on a scale that uses up the oxygen in the water. If the removal of oxygen is too great, a fish kill may result.

Any chemical applied to control a problem with nuisance organisms must be used with caution. The concentrations recommended above will affect only a portion of the life system. Large amounts of chemicals may be a danger to other life forms in the environment. If there is any doubt about the effects that treatment might have on other life systems, one should ask the responsible environmental agencies.

Tastes and odors in water can usually be removed by passing the previously filtered and chlorinated surface water through an activated carbon filter. These filters may be helpful in improving the taste of small quantities of previously treated water used for drinking or cooking purposes. They also adsorb excess chlorine. Carbon filters can be bought commercially and need regular servicing.

Carbon filters should not be expected to be a substitute for sand filtration and disinfection, however. They are not able to reliably filter raw surface water and will clog very rapidly when filtering turbid water.

Weed Control. The growth of weeds around a pond should be controlled by cutting or pulling. Before weedkillers are used, the local health department should be contacted for advice, since herbicides often contain compounds that are highly toxic to humans and animals. Algae in the pond should be controlled, particularly the blue-green types that produce scum and unpleasant odors and that, in unusual instances, may harm livestock.

PACKAGE PLANTS

Many small water treatment facilities use "package plants" which can be purchased as individual unit process modules or as a complete preassembled unit from a single manufacturer (see Figure 16). Such package plants are available from a number of manufacturers. They are most commonly supplied for filtration and removal of dissolved iron and manganese. The package plant usually includes all treatment equipment, pumps, chemical feeders, and control instrumentation. As soon as the water pipes and electrical power have been connected, the plant is ready to operate. Some package plants are fully automated.

The package plant option is frequently a good, quick choice to provide the needed treatment. Many plants can be tailored to the specific treatment needs of a particular source water. They also provide a design and equipment that have been proven to be effective and reliable in the field. However, many states require that package plant installations be formally reviewed and approved by appropriate agencies in order to certify that the plant design will meet specific requirements.

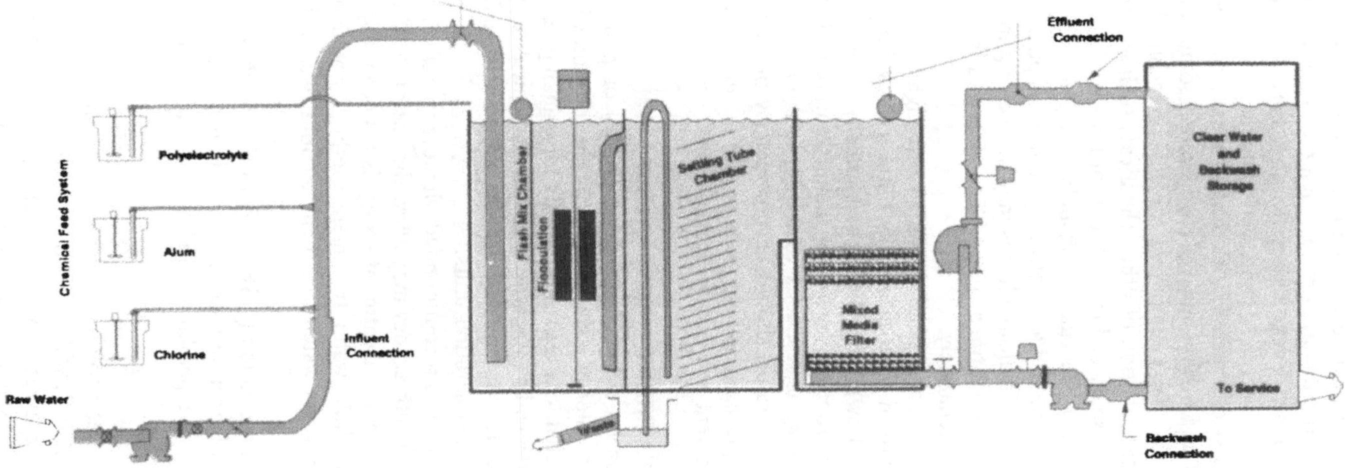

FIGURE 16. - *Package plant diagram.*

From "Technologies for Upgrading or Designing New Drinking Water Facilities", EPA (1990)

Some package plants do not operate automatically. Even those systems that are automated to some extent by instrumentation require maintenance, repairs, and process control changes by trained personnel. Such "automatic" plants are easily upset by sudden changes in source water quality.

A typical package plant is skid mounted from the factory and includes provisions for feeding a coagulant chemical and a filter aid polymer before some form of filtration. Liquid hypochlorite or chlorine gas feed equipment can also be included. Manufacturers can also make provisions for the addition of other chemicals such as carbon, potassium permanganate, soda ash, caustic soda, or lime. A typical flow configuration consists of coagulant addition, a mix chamber, flocculation chamber, settling, and filtration. The filtration step may consist of a zeolite bed at a softening facility. A pump may be included to pump finished water to a clearwell, although this may not be required if the clearwell is located at a lower level than the treatment plant. The clearwell is usually not part of the plant and must be provided separately.

Filter backwash supply pumps are usually included, although backwash water is not. They can be provided by the finished water clearwell. Filter backwashing may be initiated by filter headloss (resistance to flow), time clocks, or manually.

Most equipment manuals from package plant manufacturers include brief sections on operating principles, operating instructions and trouble-shooting guides. Start-up training, essential to the plant operator, should be provided by the manufacturer. The owner should also consider requiring follow-up visits by the manufacturer at 6-month to 1-year intervals to adjust equipment and review operations.

HOUSEHOLD WATER TREATMENT

Household treatment can be either point-of-use (POU) or point-of-entry (POE) equipment. POU devices treat water at a single tap near the location where the water is used. They are directly connected to a tap and are only for drinking and cooking water. POE treatment devices are designed to improve the microbiological or chemical quality of all water entering a household. Some POE devices are used specifically for radon removal, such as carbon filters and aeration units.

Centralized treatment of water is recommended whenever possible. However, under certain conditions, the primary health agency may allow the use of POU or POE systems for a public water supply when a central treatment system is not economically or physically practicable. Most of the household water treatment methods noted below are more completely discussed in the previous water treatment sections, which deal with central water treatment systems.

The National Sanitation Foundation (NSF) tests POE and POU household water treatment units and lists those which have passed their testing procedures. Although the NSF evaluation is voluntary, many reputable manufacturers have their equipment tested by NSF.

Water Softeners

The most common type of water softeners exchange positively charged ions (i.e., ions that cause "hardness" and rust stains), with sodium by passing water through a softening resin. After a period of time, the resin must be recharged by flushing a sodium salt solution through the softener. Sometimes, other salt solutions can be used, such as potassium salts.

As mentioned above, softening may add sodium salt to drinking water. Softening only the hot water, leaving the cold drinking water untreated, will avoid this problem. Softening may also make the water more corrosive, and possibly increase the levels of metals like lead and copper in the water. Occasional "flushing" of water at the tap will help solve the second problem.

Physical Filters

These can be made of fabric, fiber, ceramic screening, diatomaceous earth or other material. Some can remove microorganisms (cysts, larger bacteria) and very small particles (asbestos fibers); but most only remove larger particles like grit, dirt or rust. Maintenance includes monitoring the filter performance and changing cartridge filters as needed.

Activated Carbon Filters

Activated carbon filters are available in several forms: granular activated carbon (GAC), powdered activated carbon (PAC), PAC coated paper filters, and pressed carbon block filters. Activated carbon can remove many organic chemical contaminants, tastes, odors and color. Organics that are less soluble in water are removed easier than soluble ones. Activated carbon is not usually an effective system for removing microorganisms. Therefore, activated carbon filters by themselves, are not recommended for use with water which is microbiologically unsafe.

Some POE units have been used to remove radon from household water. However, because granular activated carbon (GAC) "collects" the radioactive byproducts of radon, this method of treatment (i.e., GAC) should not be used where levels of radon greater than 1,000 pCi/L are found in the source water. Aeration of household water would be more suitable for radon removal when properly installed with ventilation.

Depending on your treatment needs, either a point-of-use (POU) unit or a point-of-entry (POE) device may be used. For taste and odor reduction or organics removal, a POU unit would treat only the drinking water at the tap while a POE unit would remove contaminants from all of the water entering your home.

If proper maintenance is not followed, contaminants may be able to pass through the filter into the drinking water. As such, carbon filters have a limited life and should be replaced regularly. Bacteria can also collect and multiply on the filter's surface. Periods of non-use, especially in warm areas, promote bacterial growth which can lead to unpleasant tastes or odors in the filtered water.

Reverse Osmosis (R.O.)

A complete R.O. point-of-use system consists of an R.O. module and a storage tank. However, many home R.O. systems include one or two activated carbon filter components. The R.O. module, the heart of the system, contains a semi-permeable membrane which allows treated water to pass through it and collect in the storage tank. Other chemicals and particles are rejected and flushed to waste. R.O. is effective for removing most contaminants such as salts, metals (including lead), asbestos, nitrates, and organics.

For every four gallons of water fed into an average small household R.O. treatment unit, only about one gallon of drinking water is produced. The other three gallons of water contain the removed contaminants and is waste water. A small treated-water storage tank is used with an R.O. system because of the low pressure and flow from the treatment unit.

Water treated with the R.O. process is corrosive and might dissolve metal from plumbing or faucets. R.O. systems which use cellulose acetate membranes are not recommended for use on microbiologically unsafe water, partly because some organisms might leak through broken membranes. Bacteria can damage cellulose acetate R.O. membranes and cause them to fail if the manufacturer's recommendations on proper care and maintenance are not followed.

Ultraviolet (UV) Treatment

The primary component of a UV unit is the UV light source. The light is enclosed in a protective transparent sleeve and mounted so that the water can pass by and be exposed to the light. UV light in sufficient intensity for a long enough time can destroy bacteria and inactivate viruses. UV is not effective in controlling *Giardia lamblia* and other cysts, however.

Ultraviolet light produces no taste or odor, and treats without adding chemicals to the water. However, once the water passes through the unit, there is no disinfectant residual in the water to prevent bacteria from regrowing if the water is stored. Turbidity, dirt build-up on the transparent parts of the unit and other visible contaminants can degrade the performance of UV treatment. Regular inspection and maintenance is necessary.

Maintenance of Household Units

More sophisticated household systems can combine two or more treatment technologies so that the faults of one technology can be covered by another. However, this increases both the initial cost of purchase and the operating/maintenance costs over the life of the system.

Neglected maintenance is probably the biggest problem with home water treatment. It is important to be familiar with the maintenance requirements of each treatment unit. Some units require more maintenance than others. All should be maintained according to the manufacturer's recommendations. Some units have dealer or manufacturer maintenance contracts available to ensure proper operation over the life of the unit.

Each water supply and each water treatment unit is different. Consequently, one cannot assume that a specific unit will be the right one in every case. The National Sanitation Foundation (NSF)[3], which has a voluntary testing program for POU and POE treatment units, is a good source of information. In the absence of independent testing and listing, as by NSF, one should carefully review the data and claims for the units under consideration and get a specific written performance guarantee from the seller.

---

[3] National Sanitation Foundation, 3475 Plymouth Road, P.O. Box 1468, Ann Arbor, MI 48106.

TREATMENT WASTE DISPOSAL

Proper design and operation of a water system includes the treatment and disposal of any wastes that are produced (which may be sludges or concentrated liquids). The various waste products, and their recommended treatment and disposal, are shown as follows:

Type of waste	Sludge lagoon[1]	Sanitary Sewer	Drying beds	Evaporation ponds
Alum sludge	X	X	X	
Lime sludge	X	X	X	
Diatomic filter sludge	X	X	X	
Filter backwash water	X			
R.O./ED/EDR Concentrate		X		X
Ion exchange brine		X		X

[1] Followed by landfilling at approved site (or other benifical use)

It is also possible to discharge wastes back to their original raw water source. Care must be taken to prevent the treatment wastes from re-entering the water system, such as by discharging upstream of the water intake or into lakes where the wastes can collect and re-enter the treatment process. Discharges of this nature may require federal, state or local permits. Waste storage areas should also be located far from the water source intake to prevent contamination of the supply.

Sludge De-watering Lagoons

Lagoons are simply shallow sedimentation holding ponds. Sludge is introduced and allowed to slowly settle to the bottom while the clear supernatant (top layer), is regularly drawn off and discharged to an approved receiving water. Lagoons are usually designed to provide several years of storage, after which the sludge can be allowed to air dry and then be removed and properly landfilled or applied to some beneficial use.

Disposal to Sanitary Sewers

Disposing of sludge into a sewer merely moves the place of treatment and final disposal to a wastewater treatment plant. Considerations in this case include the limits of sewers to handle the waste, the ability of the wastewater treatment facility or septic system receiving the waste, and possible metering of the waste flows for accurate payment to the wastewater facility.

Sand Drying Beds

Liquid sludge can be spread over large, open sand beds with water removal occurring by gravity drainage, and evaporation. Dried sludge is then manually removed when it becomes too thick to allow the water to pass through, and disposed of at a landfill.

Evaporation Ponds

Ion exchange and membrane processes produce a liquid concentrate (brine) stream that may be restricted from discharge to a sewer. The concentrate must therefore be discharged to a lined pond where the liquid portion is evaporated, similar to a sludge de-watering lagoon, but with fewer solids. These ponds are, therefore, most suited for use in hot, arid climates. Other disposal methods include well injection into a salty (brackish) aquifer, brackish surface water discharge, and spray irrigation after blending with water which has fewer dissolved solids (lower TDS).

Part V

Pumping, Distribution and Storage

TYPES OF WELL PUMPS

Three types of pumps are commonly used in small and individual water distribution systems. They are: positive displacement, centrifugal, and jet. These pumps can be used in either a ground- or surface-source water system. In areas where electricity or other power (gasoline, diesel oil, solar or windmill) is available, it may be better to use a power-driven pump. Where that is not possible, a hand pump or some other manual method of supplying water must be used.

Special types of pumps with limited applications for small and individual water-supply systems include air lift pumps and hydraulic rams.

Positive Displacement Pump

The positive displacement pump delivers water at a constant rate regardless of the pressure it must overcome or the distance it must travel. These pumps are of several types.

Reciprocating Pump. This pump consists of a mechanical device which moves a plunger back and forth in a closely-fitted cylinder. The plunger is driven by the power source, and the power motion is converted from a rotating action to a back-and-forth motion by the combined work of a speed reducer, crank, and a connecting rod. The cylinder, composed of a cylinder wall, plunger, and check valve, should be located near or below the static water level to eliminate the need for priming. The pumping action begins when the water enters the cylinder. When the piston moves, the intake valve closes, and forces the water through a check valve, into the plunger. With each stroke of the plunger, water is forced toward the surface through the discharge pipe.

Helical or Spiral Rotor Pump. The helical rotor consists of a shaft with a helical (spiral) surface which rotates in a rubber sleeve. As the shaft turns, it pockets or traps the water between the shaft and the sleeve and forces it to the upper end of the sleeve.

Regenerative Turbine Pump. Other types of positive displacement pumps include the regenerative turbine type. It incorporates a rotating wheel, or "impeller", which has a series of blades or fins (sometimes called buckets) on its outer edge. The wheel is inside a stationary enclosure called a raceway or casting. Pressures several times that of pumps relying solely on centrifugal force can be developed.

Centrifugal Pump

Centrifugal pumps are pumps containing a rotating impeller mounted on a shaft turned by the power source. The rotating impeller increases the velocity of the water and discharges it into a surrounding casing shaped to slow down the flow of the water and convert the velocity to pressure.

Each impeller and its casing is called a stage. The number of stages necessary for a particular installation will be determined by the pressure needed for the operation of the water system, and the height the water must be raised from the surface of the water source to the point of use.

When more pressure is needed than can be furnished by a single stage, additional stages are used. A pump with more than one stage is called a multistage pump. In a multistage pump, water passes through each stage in succession, with an increase in pressure at each stage.

Multistage pumps commonly used in individual water systems are of the turbine and submersible types.

Turbine Pump. The vertical-drive turbine pump consists of one or more stages with the pumping unit located below the drawdown level of the water source. A vertical shaft connects the pumping assembly to a drive mechanism located above the pumping assembly. The discharge casing, pump-housing, and inlet screen are suspended from the pump base at the ground surface. The intermediate pump bearings may be lubricated by either oil or water. From a sanitary point of view, lubrication of pump bearings by water is preferred, since lubricating oil may leak and contaminate the water.

Submersible Pump. When a centrifugal pump is driven by a closely- coupled electric motor constructed for underwater operation as a single unit, it is called a submersible pump. (See Figure 17) The electrical wiring to the submersible motor must be waterproof. The electrical control should be properly grounded to minimize the possibility of shorting and damaging the entire unit. Since the pump and motor assembly are supported by the discharge pipe, the pipe should be large enough so that there is no possibility of breakage.

The turbine or submersible pump forces water directly into the water distribution system. Therefore, the pump assembly must be located below the maximum drawdown level. This type of pump can deliver water across a wide range of pressures; the only limiting factor is the size of the unit and the horsepower applied. When sand is present or anticipated in the water source, special precautions should be taken before this type of pump is used since the grinding action of the sand during pumping will shorten the life of the pump.

Jet (Ejector) Pump

Jet pumps are actually combined centrifugal and ejector pumps. A portion of the discharged water from the centrifugal pump is diverted through a nozzle and "venturi tube", which has lower pressure than the surrounding area. Therefore, water from the source (well) flows into this area of reduced pressure. The speed of the water from the nozzle pushes it through the pipe toward the surface, where the centrifugal pump can lift it by suction. The centrifugal pump then forces it into the distribution system. (See Figure 18)

Solar Photovoltaic (PV) Pump

Solar powered pump systems include a photovoltaic (PV) array (the power source), a motor and a pump. The array is made up of solar PV cells mounted to form modules which are then wired together. The solar PV cells convert sunlight directly into electricity, which drives the pump. There are several types of solar pumping systems: submersible,

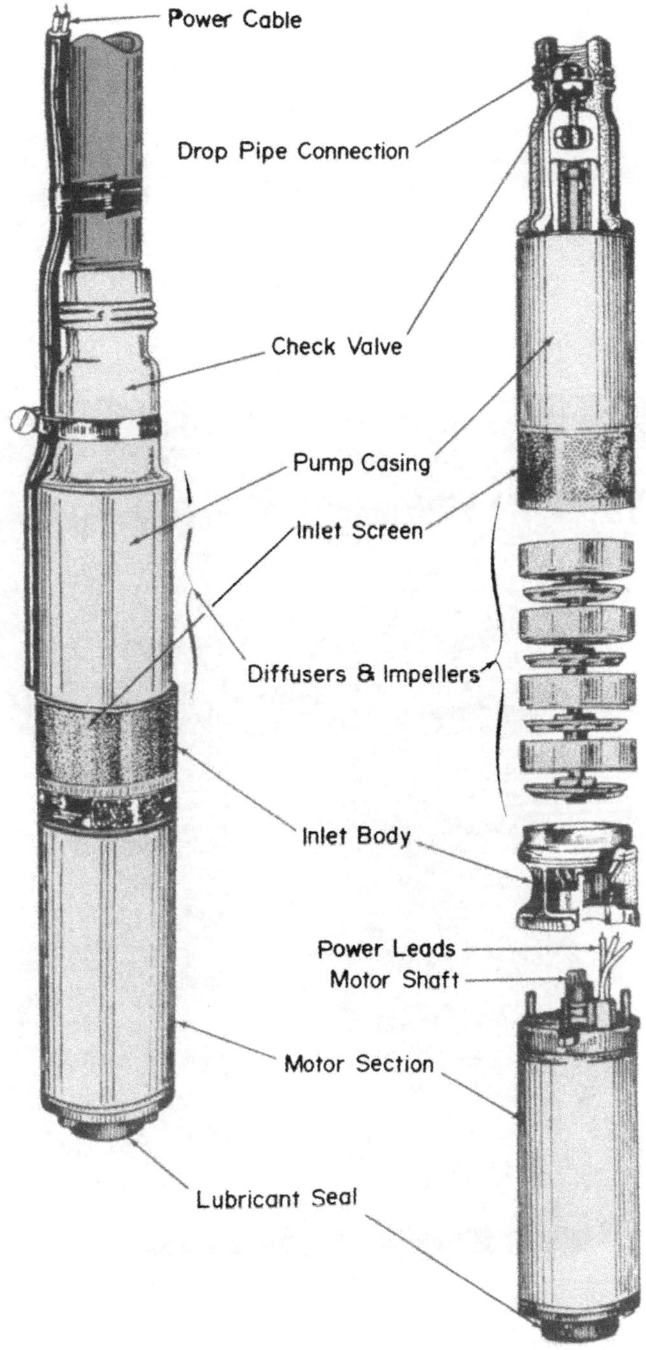

FIGURE 17. - *Exploded view of submersible pump.*

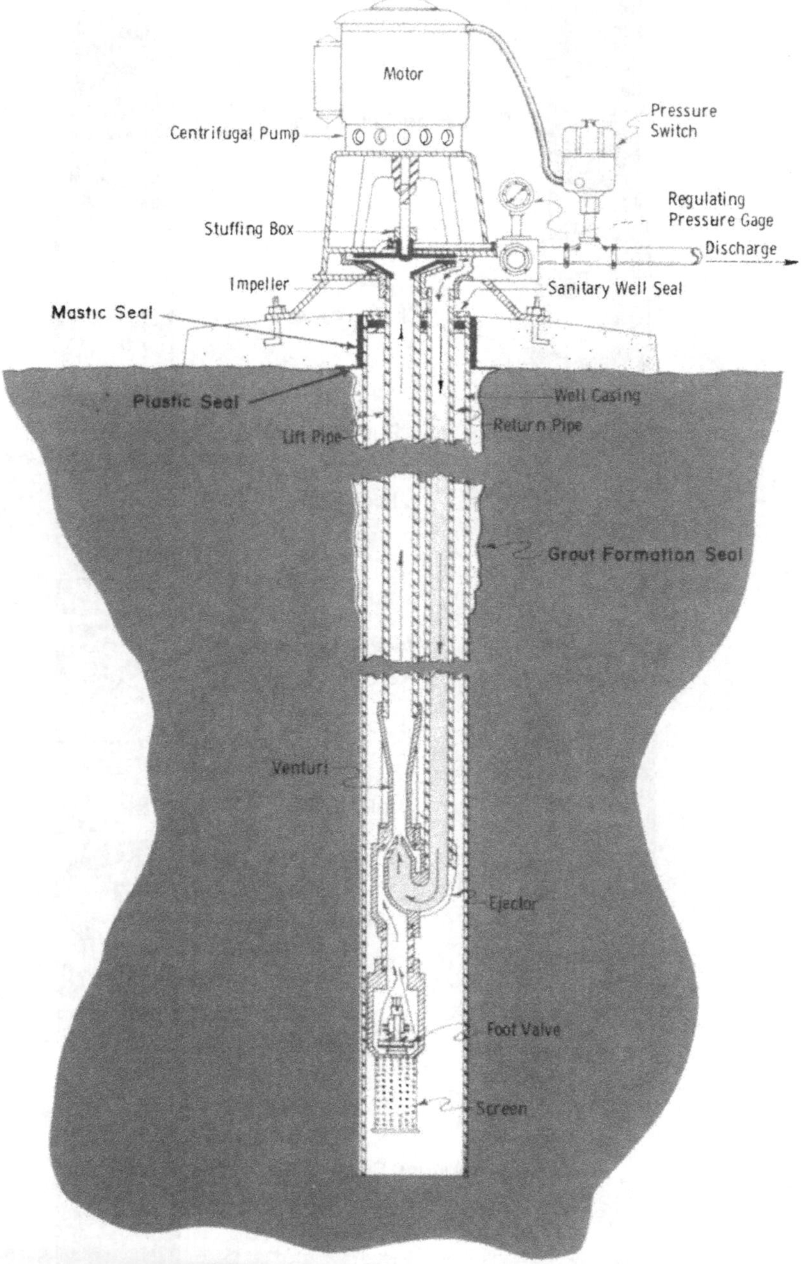

FIGURE 18. - ***"Over-the-well" jet pump installation.***

vertical turbine, floating suction and centrifugal suction pump[1]. Figure 19 shows two types of solar installations.

The most common kinds of pumps used in PV systems are self-priming centrifugal pumps, submersible centrifugal pumps, and positive-displacement (reciprocating piston) pumps.

The amount of water output for a PV pump depends upon the strength of the sunlight falling on the PV array, the total pumping pressure, and the operating temperature of the array. More intense sunlight produces more electrical output and more water. On the other hand, cloudiness reduces sunlight and water output. At times when no water can be pumped, the system needs a backup system, or battery bank for electrical power. Solar pumps require a water storage system so that water will be available to users during periods of reduced sunlight. Energy can be stored either electrically (using batteries) or hydraulically (water in storage tank).

Wind Pump

Wind pumping systems, or windmills, convert the energy in wind into mechanical or electrical energy to drive a pump (Figure 20). The energy systems are generally subdivided into two divisions: horizontal or vertical-axis machines (rotor axis), and electrical or mechanical. The most common wind pump, the horizontal-axis wind wheel, converts wind power to rotary shaft power of the axis. A windmill "head" changes the rotary motion of the axis to a vertical back-and-forth motion. This windmill "head" is

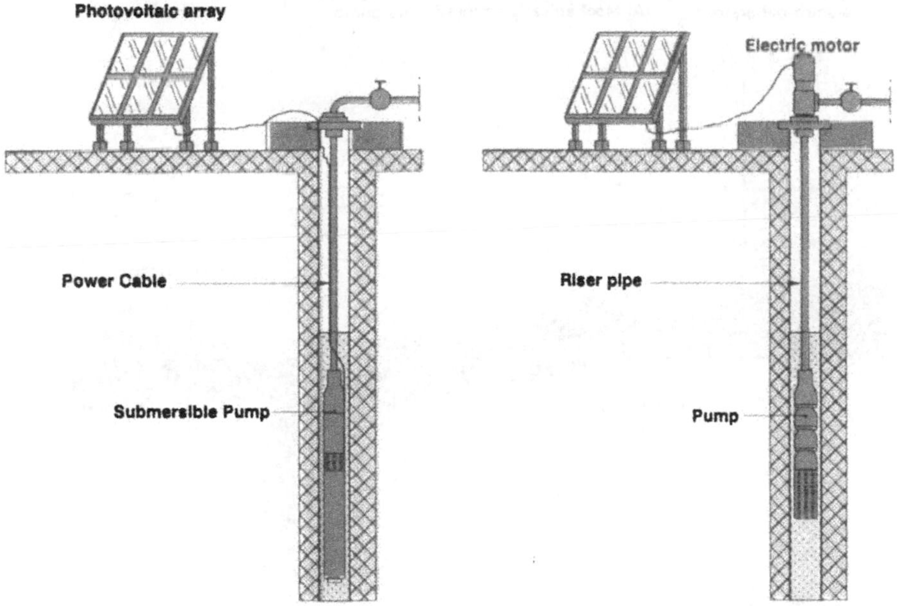

FIGURE 19. - *Typical solar pump system.*[1]

---

1 Renewable Energy Sources for Rural Water Supply, IRC International Reference Center for Community Water Supply and Sanitation, December 1986.

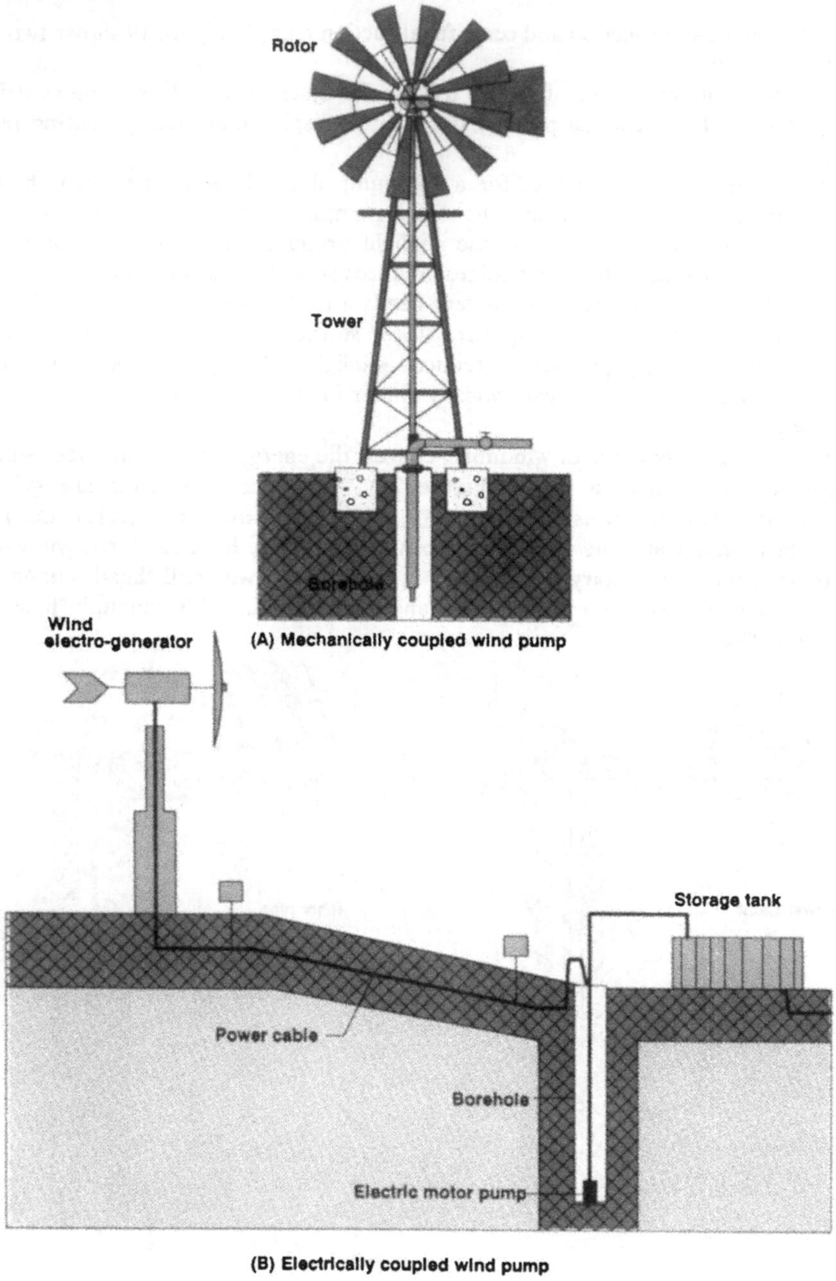

FIGURE 20. - *Typical wind powered pumps.*

From "Renewable Energy Sources for Rural Water Supply" IRC International Reference Center for Community Water Supply and Sanitation, December 1986.

mounted on a tower usually built of preformed steel. A "tail" attached to the head allows the windmill to track the changes in wind direction and stop turning during high winds with the aid of a brake. The horizontal axis mechanical system is generally used with a reciprocating piston pump. The other type of wind pump, the vertical axis design, has proven ineffective in pumping water.

The single most important factor is the wind speed at the site. There is no guarantee that a specific amount of water will be delivered over a given period of time because of variations in wind direction, frequency, and speeds. A storage tank is needed to hold enough water for three to four days to insure availability of water to users.

Hand Pump

All hand pumps depend on human power and therefore have a limited pumping rate and pressure range compared to other mechanical systems. Hand pumps (Figure 21) may be used to deliver water to the wellhead or to a storage tank. Other versions include hand pumps that deliver to stand pipes, or yard-tap distribution systems. There are four categories of hand pumps: high-lift positive displacement pumps, intermediate lift pumps, low lift pumps and suction pumps.

High-Lift Pump. The most common high-lift pumps are deep-well reciprocating piston pumps, progressive cavity pumps and deep-well diaphragm pumps.

The deep-well reciprocating piston pumps usually have underwater pump cylinders and are operated with lever-arm pump handles. They can lift water from up to about 600 feet (180 meters).

The progressive cavity pump works using a rotor turning within a fixed housing that forces water up a pipe. The deep-well diaphragm design pump uses a flexible membrane that is repeatedly stretched and relaxed mechanically to provide the pumping action. These pumps are especially effective for sandy or silty water. However, they are more complex and expensive then the reciprocating piston pump.

Intermediate-Lift/Low-Lift Pumps. The low and intermediate lift pumps are simplified versions of the high-lift pumps. Direct-action systems are suitable only for lifts of up to approximately 40 feet (12 meters) because the pumper does not have the advantage of a lever arm. These low-lift pumps are used in areas requiring less than 400 gallons per day (1.5 M^3/day).

Suction Pump. These pumps operate by creating a partial vacuum to pull the water upward. They are typically used for very low heads and lift water no more than about 22 feet (7 meters).

Other Types of Pumps

An air-lift pump (Figure 22) uses bubbles of compressed air, injected at the base of a discharge pipe, to raise water up to the surface. A fairly high air pressure (increasing with depth) is needed to inject compressed air into the submerged discharge pipe. Therefore, a great deal of power is needed to run the compressor. Advantages of this system include its simplicity, and the fact that it is not damaged by sand or silt, which are common in some wells. Also, the mechanical equipment (the compressor) is above ground, and may be used to operate other air-lift pumped wells nearby, as needed.

A hydraulic ram lifts water to a location higher than the surface or stored supply. It is a simple device that requires no external power supply, and is most suitable in hilly areas. The hydraulic ram converts pressure surges into energy, which forces water into

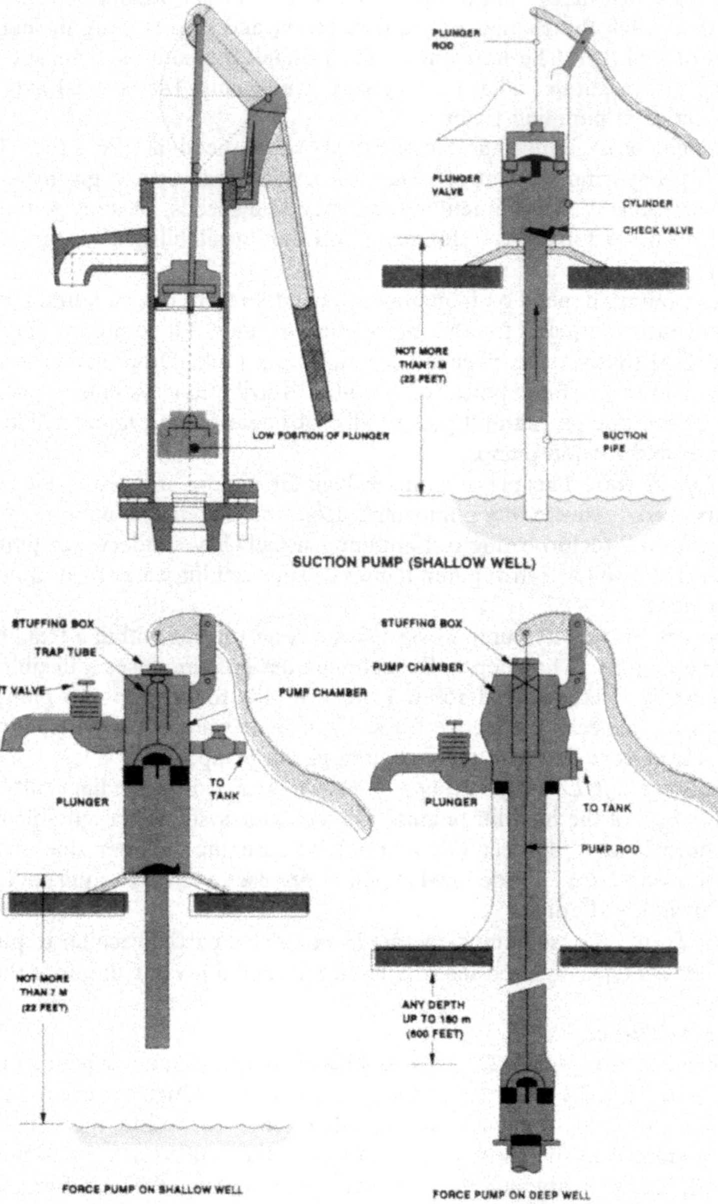

FIGURE 21. - ***Typical hand pumps.***

From International Reference Center
Technical Paper No. 18 (1987)

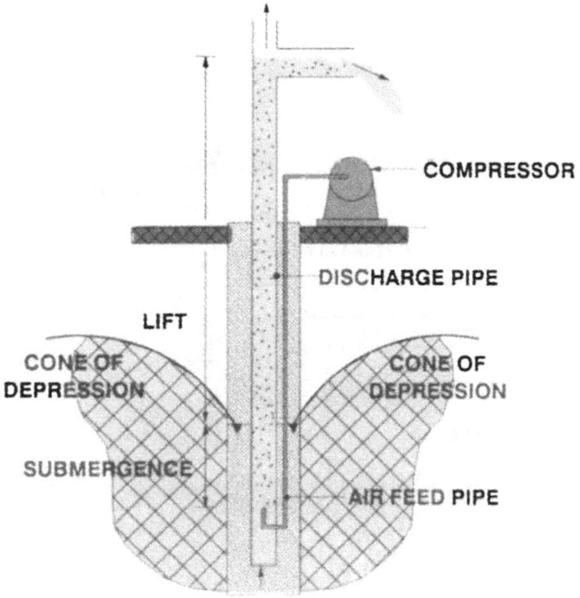

FIGURE 22. - *Typical air lift pump.*

From International Reference Center, Technical Paper No. 18;
Adapted from Hofkes and Visscher (1987)

a delivery pipe. It operates at peak efficiency if the supply pressure is about 1/3 of the delivery pressure. The parts of the hydraulic ram are shown in Figure 23.

SELECTION OF PUMPING EQUIPMENT

The type of pump for a particular installation should be chosen on the basis of the following basic considerations:

1. Well or water source capacity.
2. Daily needs and peak demand flows of the users.
3. The "usable water" in the pressure or storage tank.
4. Quality of water (presence of sand, turbidity, etc.).
5. Size and alignment of the well casing.
6. Difference in elevation between ground level and water level in the well during pumping. In other words, how high (vertically) do we have to pump the water? (Static pressure).
7. Total operating pressure of the pump at normal delivery rates, including lift and all friction losses.
8. Piping or hose length involved. How far (horizontally) do we have to pump the water?
9. Availability of electric power, or other energy source (e.g., solar, wind).
10. Ease of maintenance and availability of replacement parts.
11. First cost and economy of operation.
12. Reliability of pumping equipment.

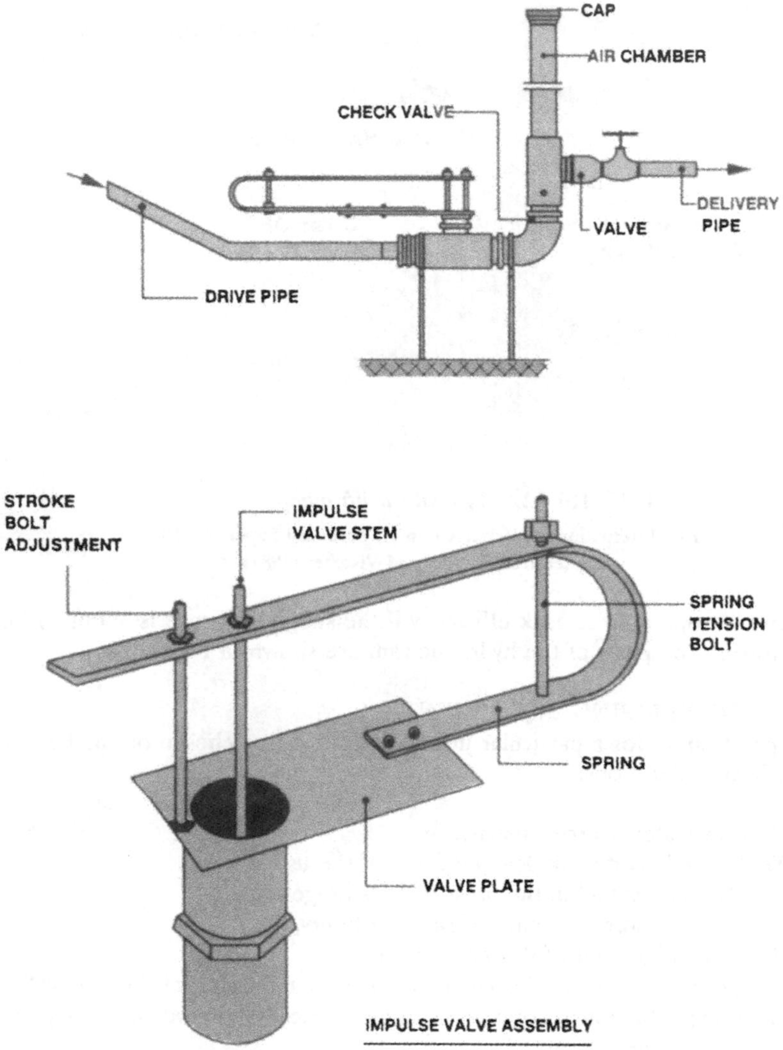

FIGURE 23. - *Typical hydraulic ram.*

From International Reference Center, Technical Paper No. 18;
Adapted from Hofkes and Visscher (1987)

In the case of solar and wind powered pumps, there are additional factors to consider. These include, for example, the average sunlight strength (in order to estimate solar array size), or estimated wind speeds at the site and at the height of the windmill (before designing a wind powered system).

The rate of water delivery required depends on both the time the pump has been operating and the total water usage between periods of pumping. Total water use can be determined from Table 3 (Page 23). The period of pump operation depends on the quantity of water on hand to meet peak demands and the storage available. If the well yield permits, a pump capable of meeting the peak demand should be used.

When the well yield is low compared to peak demand requirements, an increase in the storage capacity is required. Excessive starting and stopping will shorten the life of an electric drive motor. Therefore, the water system should be designed so that the interval between starting and stopping is as long as is possible, but not less than one minute.

Counting the number of fixtures in the home permits a ready determination of required pump capacity from Figure 24. For example, a home with kitchen sink, toilet, bathtub, wash basin, automatic clothes washer, laundry tub and two outside hose spigots, has a total of eight fixtures. Referring to the figure, it is seen that eight fixtures requires a recommended pump capacity between 9 and 11 gallons per minute. The lower value should be the minimum. The higher value might be preferred if additional fire protection is desired, or if garden irrigation or farm use is planned (see Table 9). (**Note:** This simple calculation does not take into account the possibility that low well capacity may limit the size of pump that should be installed. In this case, the system can be supplemented with additional storage to help cover periods of peak demand.

The total "operating head", or operating pressure, of a pump consists of the "lift" (vertical distance from pumping level of the water source to the pump), elevation (vertical distance from pump to highest point of water delivery), any friction losses in the pipe and fittings between the water source and pump (depends upon rate of flow, length, size, type of pipe, and on the number and type of fittings), and the discharge pressure at the pump (vertical height to which water would be raised without any outflow). For general purposes, the total head of the average system is the sum of the static head, or pressure, plus friction losses in the system. (See Figure 25)

Pumps that cannot be completely submerged during pumping use suction to raise water from the source by reducing the pressure in the pump column, creating suction. The vertical distance from the source (pumping level) to the axis of the pump is called the "suction lift", and for practical purposes cannot exceed between 15 and 25 feet (depending on the design of the pump and its altitude above sea level).

Shallow well pumps should be installed with a foot valve at the bottom of the suction line or with a check valve in the suction line in order to maintain pump prime.

The selection of a pump for any specific installation should be based on competent advice. Authorized factory representatives of pump manufacturers are among those best qualified to provide this service.

TABLE 9.- *Information on pumps.*

Type of Pump	Practical Suction Lift (1)	Usual Well-Pumping Depth	Usual Pressure Heads	Advantages	Disadvantages	Remarks
Reciprocating:						
1. Shallow well 2. Deep well	22-25 ft. 22-25 ft.	22-25 ft. Up to 600 ft.	100-200 ft. Up to 600 ft. above cylinder	1. Positive action. 2. Discharge against variable heads. 3. Pumps water containing sand and silt. 4. Especially adapted to low capacity and high lifts.	1. Pulsating discharge. 2. Subject to vibration and noise. 3. Maintenance cost may be high. 4. May cause destruction pressure if operated against closed valve.	1. Best suited for capacities of 5-25 gpm against moderate to high heads. 2. Adaptable to hand operation. 3. Can be installed in very small diameter wells (2" casing). 4. Pump must be set directly over well (deep well only).
Centrifugal:						
1. Shallow well a. Straight centrifugal (single stage)	20 ft. max.	10-20 ft.	100-150 ft.	1. Smooth, even flow. 2. Pumps water containing sand and silt. 3. Pressure on system is even and free from stock. 4. Low-starting torque. 5. Usually reliable and good service life.	1. Looses prime easily. 2. Efficiency depends on operating under design heads and speed.	1. Very efficient pump for capacities above 60 gpm and heads up to about 150 ft.
b. Regenerative vane turbine type (single impeller)	28 ft. max.	28 ft.	100-200 ft.	1. Same as straight centrifugal except not suitable for pumping water containing sand or silt. 2. They are self-priming.	1. Same as straight centrifugal except maintains priming easily.	1. Reduction in pressure with increased capacity not as severe as straight centrifugal.
2. Deep well a. Vertical line shaft turbine (multi- stage)	Impellers submerged	50-300 ft.	100-800 ft.	1. Same as shallow well turbine. 2. All electrical components are accessible, above ground.	1. Efficiency depends on operating under design head and speed. 2. Requires straight well large enough for turbine bowls and housing.	

					3. Lubrication and alignment of shaft critical. 4. Abrasion from sand.	
b. Submersible turbine (multi-stage)	Pump and motor submerged	50-400 ft.	50-400 ft.	1. Same as shallow well turbine. 2. East to frost-proof installation. 3. Short pump shaft to motor 4. Quiet operation. 5. Well straightness not critical.	1. Repair to motor or pump requires pulling from well. 2. Sealing of electrical equipment from water vapor critical. 3. Abrasion from sand.	1. 3,500 rpm models, while popular because of smaller diameters of greater capacities, are more vulnerable to wear and failure from sand and other causes.
Jet:						
1. Shallow well	15-20 ft. below ejector	Up to 15-20 ft below ejector	80-150 ft.	1. High capacity at low heads. 2. Simple in operation. 3. Does not have to be installed over the well. 4. No moving parts in the well.		
2. Deep well	15-20 ft. below ejector	25-120 ft. 200 ft. max.	80-150 ft.	1. Same as shallow well jet. 2. Well straightness not critical.	1. Same as shallow well jet. 2. Lower efficiency, especially at greater lifts.	1. The amount of water returned to ejector increases with increased lift--50% of total water pumped a 50-ft. lift and 75% a 100-ft. lift.
Rotary:						
1. Shallow well (gear type)	22 ft.	22 ft.	50-250 ft.	1. Positive action. 2. Discharge constant under variable heads. 3. Efficient operation.	1. Subject to rapid wear if water contains sand or silt. 2. Wear of gears reduces efficiency.	
2. Deep well (helical rotary type)	Usually submerged	50-500 ft.	100-500 ft.	1. Same as shallow well rotary. 2. Only one moving pump device in well.	1. Same as shallow well rotary except no gear wear.	1. A cutless rubber stator increases life of pump. Flexible drive coupling has been weak point in pump. Best adapted for low capacity and high heads.

(1) Practical suction lift at sea levl. Reduce life 1 foot for each 1,000 ft. above sea level.

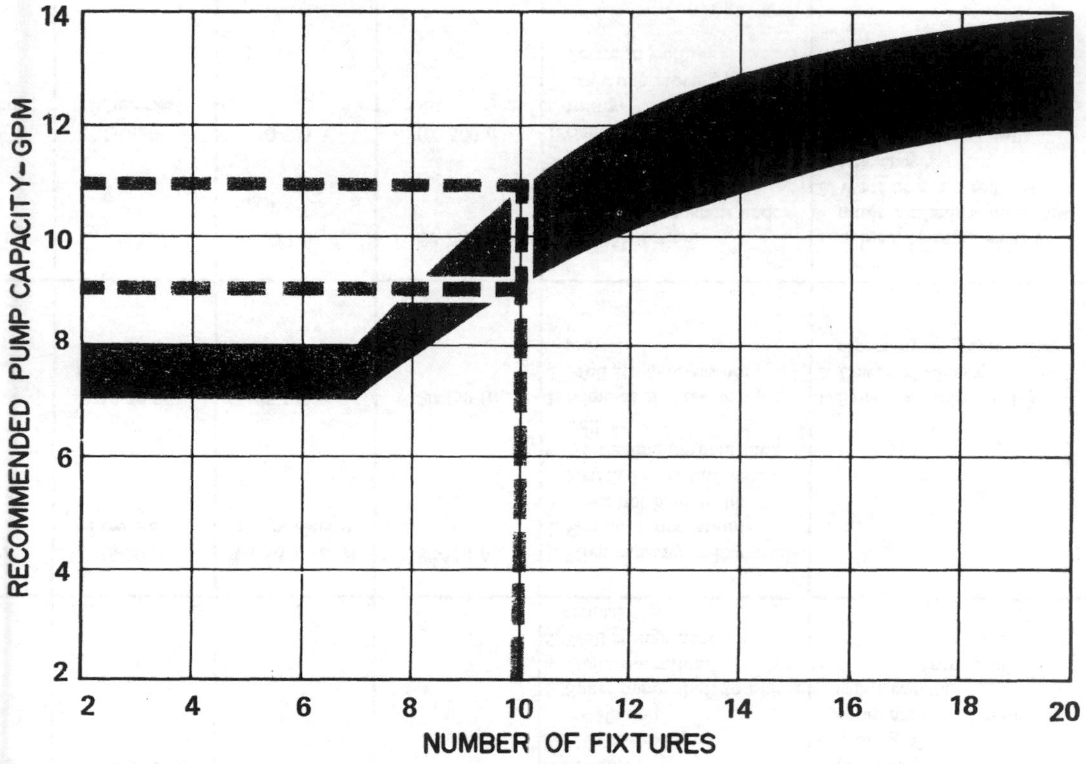

FIGURE 24. - *Determining recommended pump capacity.*

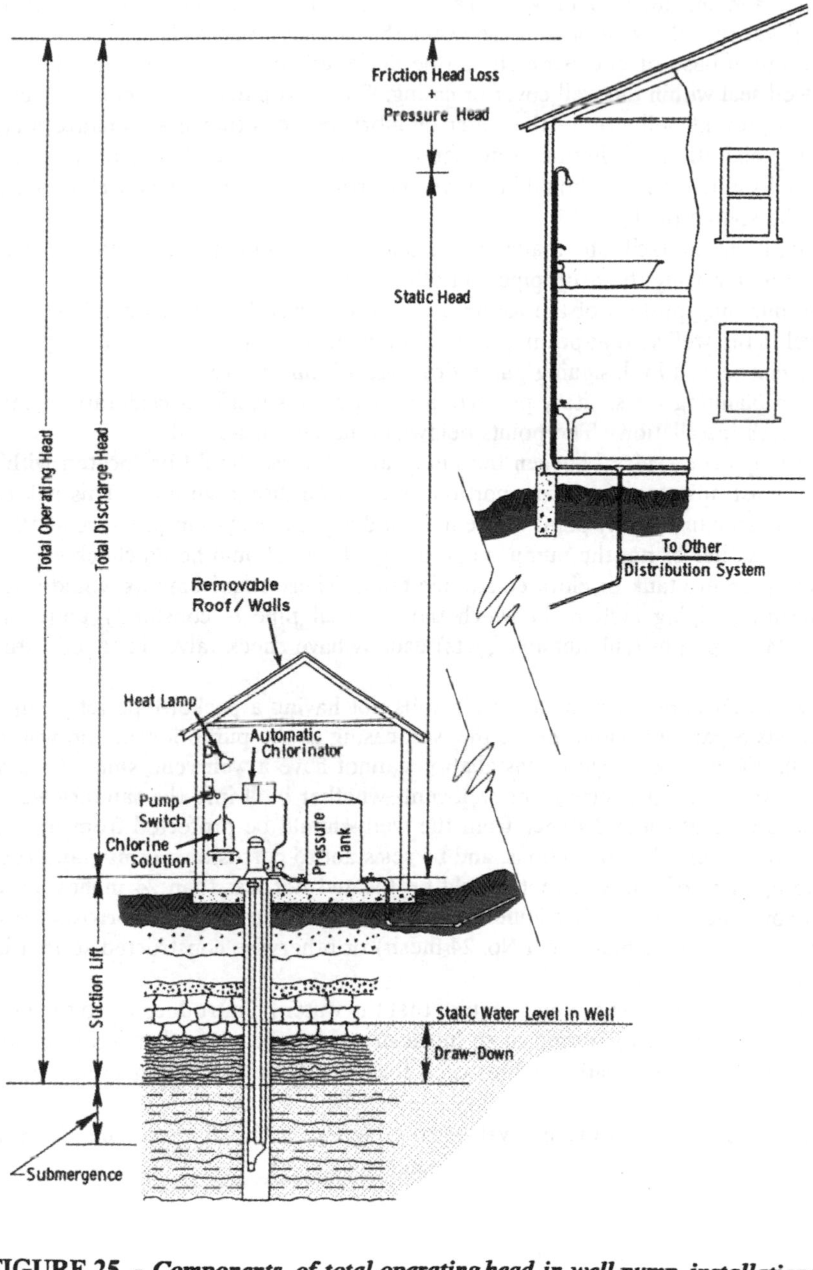

FIGURE 25. - ***Components of total operating head in well pump installations.***

SANITARY PROTECTION OF PUMPING FACILITIES

The pump equipment for either power-driven or manual systems should be constructed and installed to prevent the entrance of contamination or objectionable material either into the well or into the water that is being pumped.

The pump base or enclosure should be designed so that it is possible to install a sanitary well seal within the well cover or casing. The pump head or enclosure should be designed to prevent pollution of the water by lubricants or other maintenance materials used during operation. Pollution from hand contact, dust, rain, birds, flies, rodents or animals, and similar sources should be prevented from reaching the water chamber of the pump or the source of supply.

Design should facilitate maintenance and repair, and include enough overhead clearance for removing the drop pipe and other accessories.

The pumping portion of the assembly should be installed near or below the static water level in the well so that priming will not be necessary. If necessary, frost protection should be considered in designing pump drainage within the well.

When planning for sanitary protection of a pump, specific considerations must be made for each installation. The points below should be considered.

The only check valve between the pump and storage should be located within the well, or at least upstream from any portion of a buried discharge line. This will ensure that the discharge line at any point will remain under positive system pressure at all points of contact--whether or not the pump is operating. There should be no check valve at the inlet to the pressure tank or elevated storage tank. These requirements would not apply to a concentric piping system, in which the external pipe is constantly under system pressure. Many pumps (submersibles, jets) usually have check valves installed within the well.

A well vent is recommended on all wells not having a packer-type jet pump. The vent prevents a partial vacuum inside the well casing as the pump lowers the water level in the well. (The packer-type jet installation cannot have a well vent, since the casing is under positive system pressure.) The well vent--whether built into the sanitary well cover or routed to a point some distance from the well--should be protected from mechanical damage, have watertight connections, and be resistant to corrosion, vermin, and rodents.

The opening of the well vent should be located not less than 24 inches above the highest known flood level. It should be screened with durable and corrosion-resistant materials (bronze or stainless steel No. 24 mesh) or otherwise constructed so that insects and vermin are kept out.

Certain types of power pumps require that the water be introduced into the pumping system, either to prime the pump or to lubricate rubber bearings that have become dry while the pump was inoperative. Water used for priming or lubricating should be free of contamination.

It is a good idea to provide a water-sampling tap on the discharge line from power pumps.

INSTALLATION OF PUMPING EQUIPMENT

Where and how the pump and power unit are mounted depend primarily on the type of pump employed. The vertical turbine centrifugal pump, with the power source located directly over the well and the pumping assembly submerged in the well, is gradually being replaced by the submersible unit, where both the power unit (electric motor) and the pump are submerged. Similarly, the jet pump is gradually giving way to the submersible pump for deeper installations because of the latter's superior performance and better operating economy.

Vertical Turbine Pump

In the vertical turbine pump installation, the power unit (usually an electric motor) is installed directly over the well casing. The pump portion is submerged within the well, and the two are connected by a shaft in the pump column. The pump column supports the bearing system for the drive shaft and brings the pumped water to the surface. (See Figure 26)

Since the long shaft must rotate at high speed (1,800 to 3,600 rpm), correct alignment of the motor, shaft, and pump is vital to good performance and long life of the equipment. There are two main points to consider in installing the pump correctly. The first is correct and stable positioning of the power unit. The second is the straightness or vertical positioning of the pump column within the well.

Since concrete slabs tend to deteriorate, settle, or crack from weight and vibration, it is usually better to attach the discharge head to the well casing. Figure 26 shows one way to accomplish this. For smoothest operation and minimum wear, the plate (and discharge head) should be mounted perpendicular to the axis of the pump column as pump and column hang in the well. If the casing is perfectly vertical, the pump column axis and the well axis coincide, and a perfect installation results. It sometimes happens, though, that the well is not perfectly vertical. In this case, it is necessary to adjust the position of the plate so that the axis of the pump column is as close as possible to the axis of the well. If there is enough room inside the casing (and this is one of the reasons for installing larger casing), there is a better chance that the pump and column will be able to hang perfectly vertical--or at least be able to operate smoothly. Once the correct position of the plate is determined, it is welded to the well casing. The discharge head is then bolted securely to the support plate.

As explained under "Sanitary Construction of Wells" on Page 58, sanitary well seals or covers are available to seal the well casing against contamination. However, some designs make it difficult or impossible to measure water levels within the well. This deficiency should be corrected by welding an access pipe to the side of the casing, to permit insertion of a water-level measuring device. First, a hole is cut in the casing at a point far enough below the top to permit clear access past the discharge head of the pump. The angle between the access pipe and the casing should be small enough to allow free entry of the measuring line. The minimum inside diameter of the pipe should be 3/4 inch, and larger if possible. Before welding the pipe in place, any sharp edges around the hole through the casing should be filed smooth so that the measuring device will slide freely through the hole without catching or becoming scratched. An angle of one unit horizontal to four units vertical provides good access.

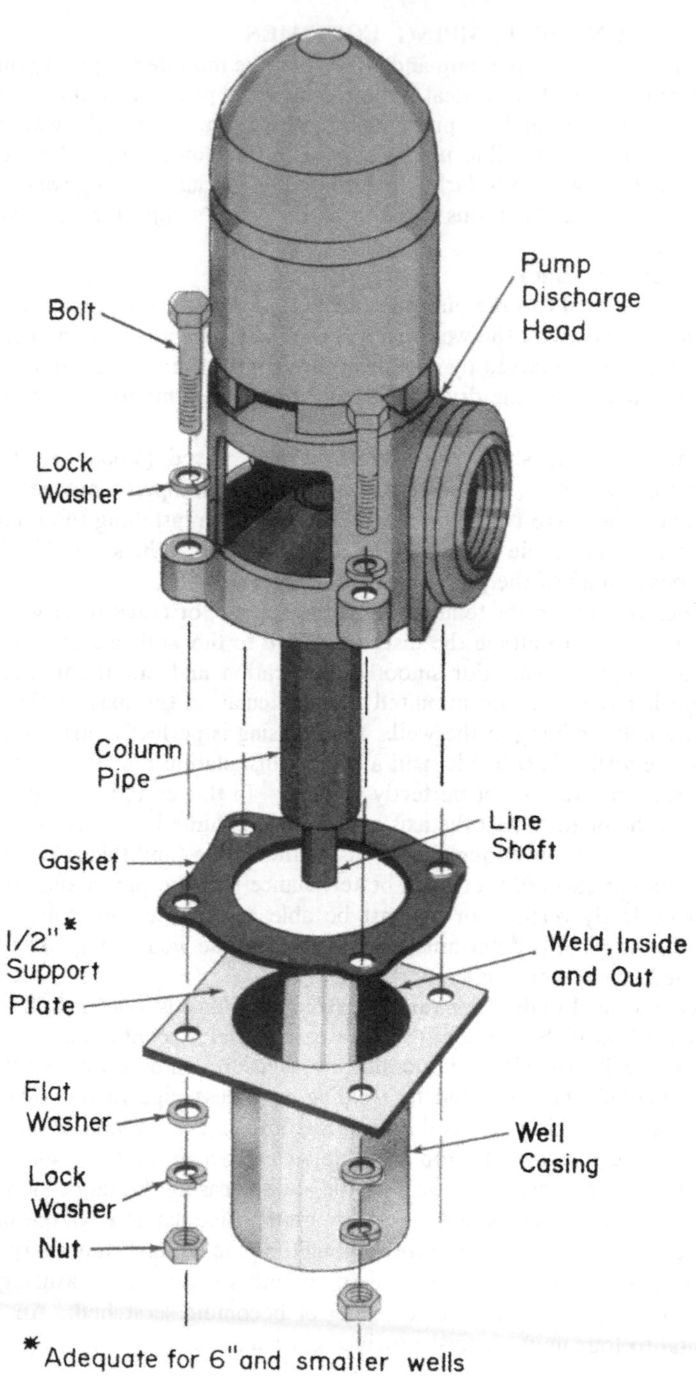

FIGURE 26. - ***Vertical (line shaft) turbine pump mounted on well case.***

It is recommended that all wells be equipped with access pipes because of the ease of introducing and removing measuring devices, and because the pipe permits chemical treatment of the well without removing the sanitary well seal and pump.

The welding around the access pipe should be at least as thick and resistant to corrosion as the well casing itself. This is especially important if the connection will be located below the ground surface.

Submersible Pump

This pump performs well in casings that are too crooked for vertical turbine pumps, because all moving parts of the submersible pump are located in one unit within the well. A problem could arise if there is little space between the inside casing and outside of the pump--the pump might stick in the well casing or be damaged during installation. If there is any doubt about whether there is enough space, a "dummy" piece of pipe whose dimensions are slightly greater than those of the pump should first be run through the casing to make sure that the pump will pass freely to the desired depth.

The entire weight of the pump, cable, drop pipe, column of water within the pipe, and reaction load when pumping must be supported by the drop pipe itself. It is important, therefore, that the drop pipe and couplings be of good quality galvanized steel and of standard weight. Cast-iron fittings should not be used where they must support pumps and pump columns.

The entire load of submersible pumping equipment is usually suspended from the sanitary well seal or cover. An exception to this would be the "pitless" installation.

Jet Pump

Jet pumps may be installed directly over the well, or alongside it. Since there are no moving parts in the well, straightness and vertical alignment do not affect the jet pump's performance. The equipment in the well is relatively lightweight, being mostly pipe (often plastic), so that loads are supported easily by the sanitary well seal. There are also a number of good "pitless adapter" and "pitless unit" designs for both single and double pipe jet systems.

ALTERNATE ENERGY SOURCES AND PUMPS

Solar Photovoltaic (PV) Pump

To decide whether a site is suitable for solar pumps, a record of daily irradiation (solar intensity) for at least one year (preferably two) is helpful. The design of a solar pump system cannot be based on average daily irradiation, because not enough water would be pumped in months of below-average solar intensity. The month of lowest irradiation should be used for designing a solar pump.

Solar photovoltaic (PV) systems use photovoltaic cells which convert solar irradiation directly into electricity for pump operation. Several pumping systems of this type are reliable, and available from a number of manufacturers[2]. The main parts of a PV pumping system are:

- Array of photovoltaic cells which convert solar irradiation into direct current (dc) electricity
- Motor and pump
- Control system
- Water storage and distribution system

Equipment can be added to improve the performance of PV pumps such as tracking mechanisms, solar concentrating devices and batteries. While tracking mechanisms and solar concentrators maximize the use of available irradiation, batteries can help provide a backup supply of electricity from the varying supply of the PV array. The batteries can also store electrical energy for use during periods when the array is not producing enough. However, batteries have several drawbacks, such as heavy power loss, lack of reliability, need for regular maintenance, and a useful life which is shorter than the rest of the solar pumping system.

The reason why PV pump systems are not more widely used is that they require a large initial investment. However, analysis of both initial equipment and operating costs indicates that PV pumps are likely to be cost-effective in areas of high levels of solar irradiation, and may become more attractive as increased production of PV cells results in lower prices.

Wind Pump

Wind pumps use a rotor to either directly drive a pump or to drive an electro-generator (which produces the electricity to operate a pump) (see Figure 20). While the mechanically driven pump requires a large rotor to capture enough wind to produce useful power, the electro-generator uses a smaller rotor, designed for reaching a high speed of rotation.

A slow running, multi-bladed rotor for mechanically driven pumps has to be of sturdy construction because of the considerable load and force created by the rotor. Rotary pumps are normally driven by gears on pulleys while a reciprocating-plunger pump often uses a crank shaft coupling. Generally, a rotary pump will put a more consistent load on the wind rotor than a plunger pump, because the power required to run a plunger pump changes constantly during the pumping cycle. In contrast, a rotary pump imposes a relatively constant and continuous load on a rotor.

A two or three-bladed rotor is used to drive the electro-generator through a set-up gear unit to get to the required high operating rotational speed. If the generator produces three-phase alternating current, this power can be supplied directly to a standard submersible motor/pump set. The maximum power rating of the electric motor should not be more than about 40 percent of the generator rating.

Wind pumps should be built using a sturdy tower that can withstand the large forces exerted on it by the rotor, especially in high winds. In addition, every wind pump should

---

2 Renewable Energy Sources for Rural Water Supply, IRC International Reference Center for Community Water Supply and Sanitation, December 1986.

have a control device to protect it from rotating too fast and being damaged in high winds. Some of these control devices include brakes, which limit rotational speed, and "pull out" devices, which turn the rotor away from high winds by changing its direction (under normal operating conditions, the vane is perpendicular to the wind, but in high winds, the force on it makes the rotor turn away from the wind). Another control device disturbs the air stream on the rotor, thereby lowering the pressure on the rotor.

In mostly flat areas with few trees or buildings, site selection is not vital. In mountainous areas or places where obstacles may block the flow of wind, differences in surface roughness and obstacles must be taken into account when estimating wind speeds for the site. Generally, the windmill tower should be tall enough so the rotor is at least 15 feet (5 meters) off the ground and that the tower stands above any obstruction within a radius of 400 feet (125 meters).

Occasionally, one needs to lubricate the bearings, tighten bolts, and make minor adjustments and repairs. A complete overhaul is normally needed every couple of years, while cup seals and other moving pump parts may need more frequent maintenance.

Hand Pump

When properly maintained, hand pumps can be the least expensive and most reliable technology for many uses. Over 50 makes and models of hand pumps are available worldwide. As mentioned before, hand pumps can be divided into four categories:

- High lift--positive-displacement pumps able to pump from depths of up to 150 feet
- Intermediate lift--for lifts of up to 80 feet.
- Low lift--up to 40 feet
- Suction pumps--up to 22 feet

Capacities of hand pumps varies from 1.5 to 5 gallons per minute depending on the pump type, how deep the water is and who is operating the pump. A hand pump may provide up to 1,200 gallons of water per day for a family or very small community.

The pump heads on most force pumps are designed with a "stuffing box" around the pump rod to protect against contamination. Lift pumps with slotted pump head tops are open to contamination and should not be used. The pump spout should be closed and directed downward.

The pump base should be designed to serve a two-fold purpose: first, to support the pump on the well cover or casing top; and second, to protect the well opening or casing top from contaminated water or other harmful material. The base should be a solid, one-piece, recessed type, cast with or threaded to the pump column or stand. It should have enough diameter and depth to permit a 6-inch well casing to extend at least 1 inch above the surface upon which the pump base is to rest. Using a flanged sleeve embedded in the concrete well cover or a flange threaded or clamped on the top of the casing to form a support for the pump base is recommended. Gaskets should be used to insure tight closure.

Regular preventative maintenance for hand pumps (such as tightening nuts and bolts and keeping the area clean), can be performed by the owner. However, when repairs are required for below-ground parts, an experienced mechanic should be hired. Some below-ground parts, that may fail include the cylinder, cylinder seals, drop pipe and pump rods.

The expected lifetime of hand pumps is difficult to predict since it depends on the amount of use and preventive maintenance effort. Typically, one can expect anywhere from 5 to 10 years of operation.

PUMPHOUSING AND APPURTENANCES

A pumphouse installed above the surface of the ground should be used. (See Figure 27) The pumproom floor should be watertight, preferably concrete, and should slope uniformly away in all directions from the well casing or pipesleeve. It should be unnecessary to use an underground discharge connection if the pumphouse is insulated and heated. For individual pumphouses in rural areas, two 60-watt light bulbs, a thermostatically controlled electric heater, or a heating cable will generally provide adequate protection as long as the pumphouse is properly insulated.

In areas where power failures may occur, an emergency, gasoline-driven power supply or pump should be considered. A natural disaster, such as a severe storm, hurricane, tornado, blizzard, or flood, may cut off power for hours or even days. A gasoline power-driven electrical unit could supply enough power for the pump, basic lighting, refrigeration, and other emergency needs.

Lightning Protection

Voltage and current surges produced in powerlines by lightning are a serious threat to electric motors. The high voltage can easily pierce and burn the insulation between motor windings and the motor frame. The submersible pump motor is somewhat more vulnerable to this kind of damage because it is submerged in ground water, the natural "ground" sought by the lightning discharge. Actual failure of the motor may be immediate, or it may be delayed for weeks or months.

There are simple lightning arresters available to protect motors and appliances from "near miss" lightning strikes. (They are seldom effective against direct hits.) The two types available are the valve type and the expulsion type. The valve type is preferred because its "sparkover" voltage remains constant with repeated operation.

Just as important as selecting a good arrester is installing it properly. The device must be installed according to instructions from the manufacturer and connected to a good ground. In the case of submersible pumps, a good ground connection can be achieved by connecting the ground terminal of the arrester to the submersible pump motor frame using a No. 12 stranded bare copper wire. The low resistance of the wire (1 ohm or less) reduces the voltage surge reaching the motor windings to a level that it can withstand.

If steel well casing extends below the watertable, the ground can be made even better by also connecting the bare copper wire to the well casing.

IMPORTANT NOTE: Connecting the ground terminal of the arrester to a copper rod driven into the ground does not satisfy grounding requirements. Similarly, if a steel casing that does not reach the ground water is relied upon, the arrester may be ineffective.

Additional advice on the location and installation of lightning arresters can be gotten from the power company serving the area.

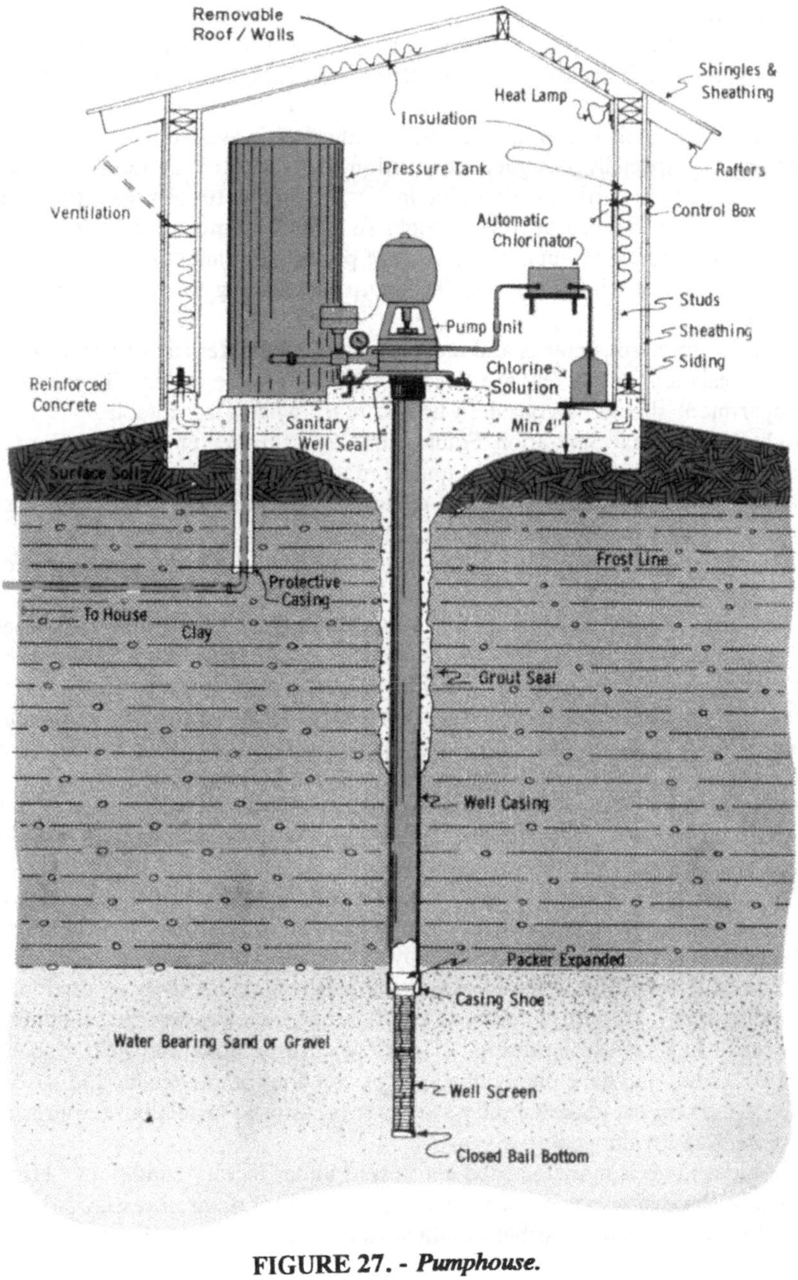

FIGURE 27. - ***Pumphouse.***

Pitless Units and Adapters

Because of the pollution hazards involved, a well pit to house the pumping equipment or to permit access to the top of the well is not recommended. Some states prohibit its use.

A commercial unit known as the "pitless adapter" is available to eliminate well pit construction. A specially designed connection between the underground horizontal discharge pipe and the vertical casing pipe makes it possible to terminate the permanent, watertight casing of the well at a safe height (8 inches or more) above the final grade level. The underground section of the discharge pipe is permanently installed and it is not necessary to disturb it when repairing the pump or cleaning the well. (See Figures 28 through 31)

There are numerous makes and models of pitless adapters and units available. Not all are well-designed, and a few are not acceptable to some states. The state or local health department should be consulted first to learn what is acceptable.

Both the National Sanitation Foundation[3] and the Water Systems Council[4] have adopted criteria intended to assure that quality materials and workmanship are used in the manufacture and installation of these devices. Unfortunately, the safety of these installations is highly dependent on the quality of workmanship used during their installation on the site. For this reason, additional precautions and suggestions are offered here.

There are two general types of pitless installations. One, the "pitless adapter," requires cutting a hole in the side of the casing at a certain depth below the ground surface (usually below the frost line). A fitting to accommodate the discharge line from the pump is then inserted into the hole and attached. Its design varies depending on whether it is for a pressure line alone or for both pressure and suction lines (a two-pipe jet pump system with pump mounted away from well). The other part of the adapter, mounted inside the well, supports the pumping components that are suspended in the well. Watertight connection is accomplished by a system of rubber seals, tightened down by clamps or by the weight of the equipment itself.

The second type, the "pitless unit", requires cutting off the well casing at the required depth and mounting on it factory assembled unit with all necessary attachments.

Regardless of the type of device employed, certain problems arise, calling for special care. Some of these are described below, with suggestions for their correction.

Welding below ground, in cramped quarters and under all-weather conditions, often does not result in good workmanship. If welding must be done, the welder should be an expert pipe welder, and have plenty of room for freedom of movement and ease of visual inspection. A clamp-on, gasketed pitless adapter is easier to install, but requires a smooth and clean surface for the gasket.

The pitless unit is manufactured and tested under factory conditions. However, its attachment to the casing may present special problems. If the well casing is threaded and coupled (T&C), it may be possible to adjust the height of one of the joints so that it is

---

3 **National Sanitation Foundation, 3475 Plymouth Road, Ann Arbor, Michigan 48105.**

4 **Water Systems Council, 600 S. Federal Street, Suite 400, Chicago, Illinois 60605.**

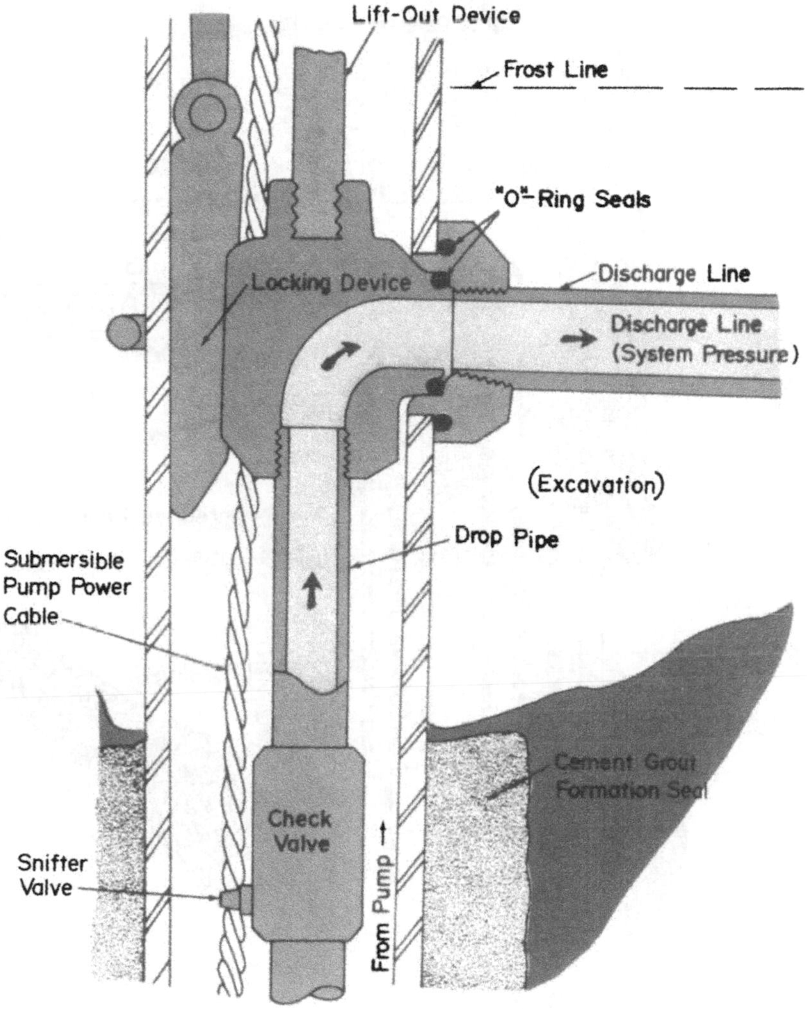

FIGURE 28. - *Clamp-on pitless adapter for submersible pump installation.*

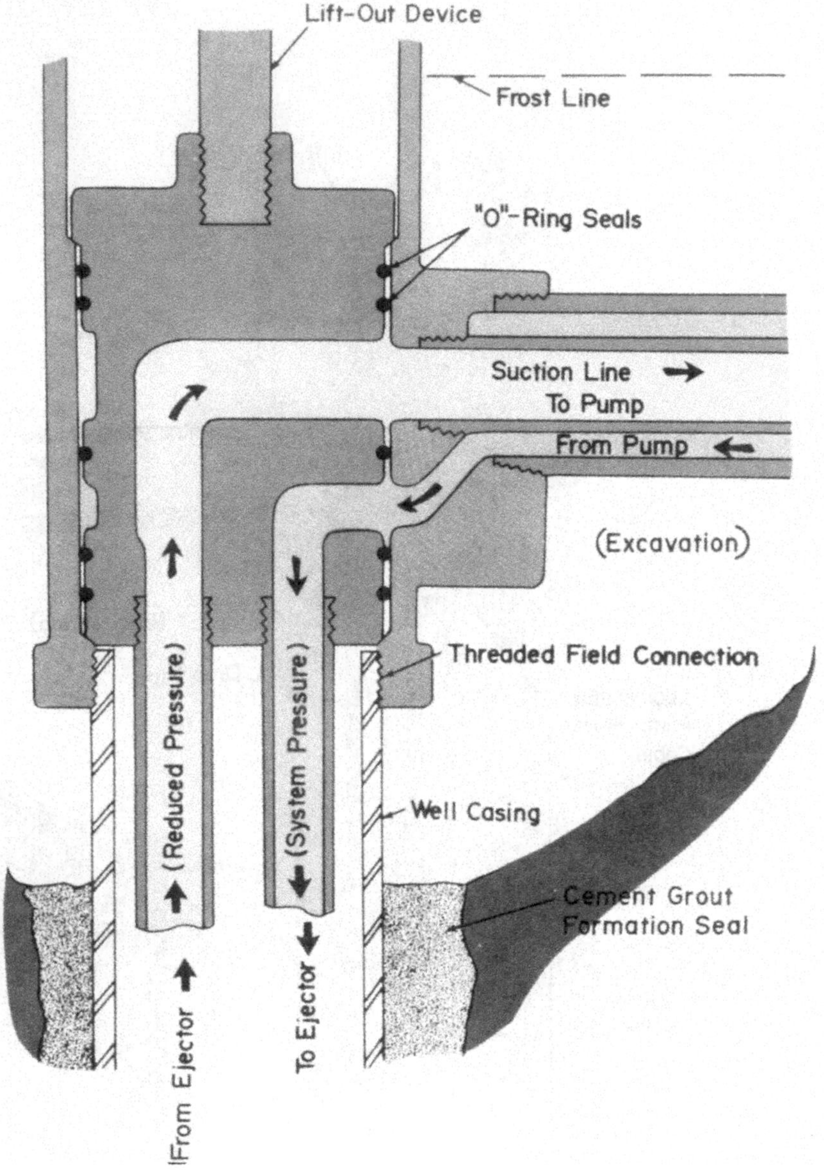

FIGURE 29. - ***Pitless unit with concentric external piping for jet pump installation.***

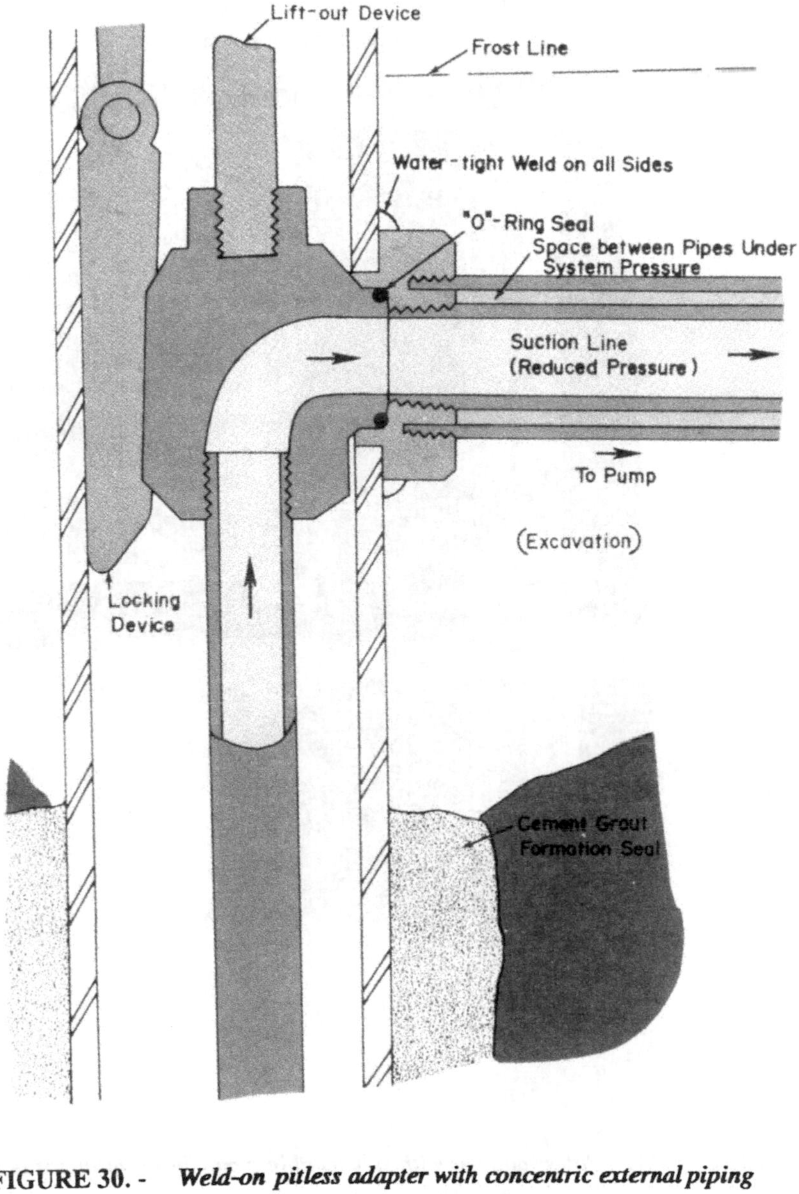

FIGURE 30. - ***Weld-on pitless adapter with concentric external piping for "shallow well" pump installation.***

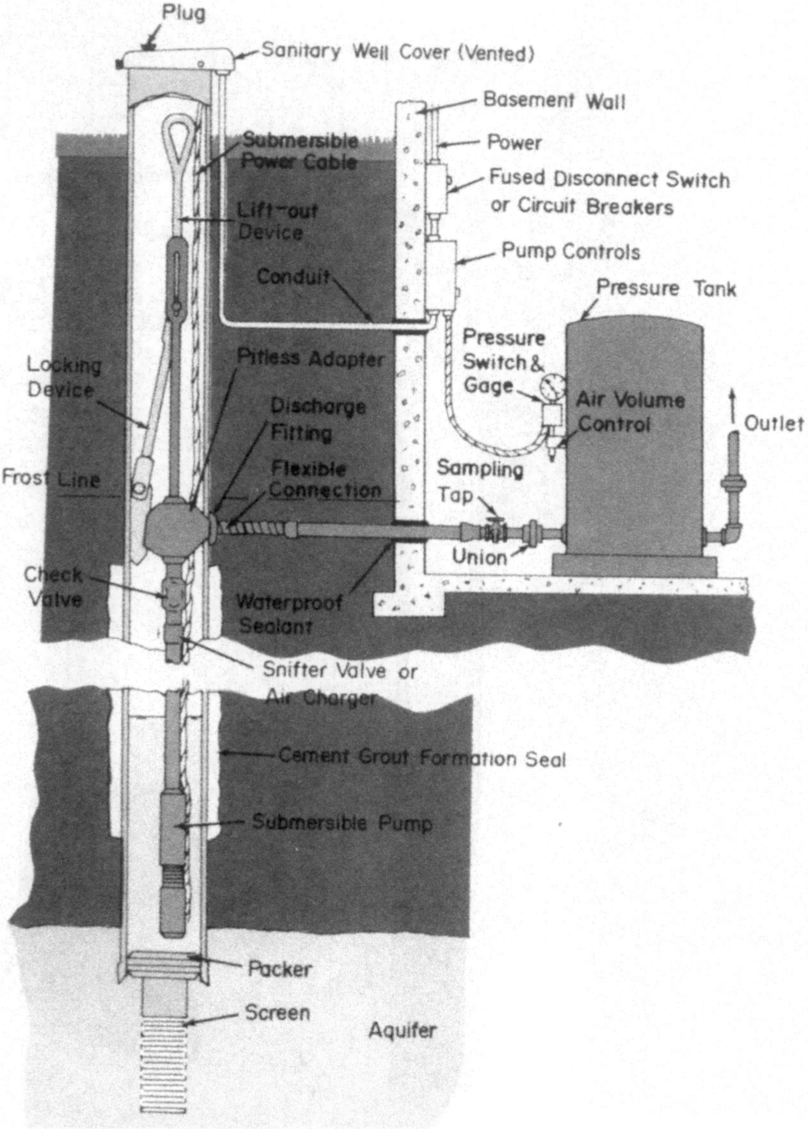

FIGURE 31. - ***Pitless adapter with submersible pump installation for basement storage.***

about right for the attachment of the unit. If the height cannot be adjusted, or if welded joints have been made, the casing must be cut off at the proper depth below ground and then threaded.

Power-driven pipe-threading machines can be used to thread casing "in place" in sizes up to and including four inches. Between ten and twelve full threads should be cut on the casing to make a good, strong joint. The threads should be of good quality, cut with dies in good condition.

When it is necessary to weld, the first requirement is that the casing be cut off squarely. This cut can be made by working inside the casing using special casing-cutting tools, or by "burning" with an acetylene torch from outside the casing. If the torch method is used, it is better to use a jig that attaches to the casing, supporting and guiding the torch as the casing is burned off.

A competent welder should be able to make a strong weld if given enough room to work in. It is not easy to get a watertight joint under these conditions. Two or three "passes" around the pipe should be made, following recommended procedures for pressure pipe welding. The final welded connection should be at least as thick, as strong, and as resistant to corrosion as the well casing itself.

Clamps and gaskets are used to attach both adapters and units. These devices have been criticized by some health departments because of their structural weakness compared with other connections. It is feared by some that the joint is more easily broken or caused to leak by mechanical damage, or by frost-heave acting on the casing or the well slab.[5]

A watertight joint requires good contact between the gasket and the surfaces it seals off. Corrosion-resistant, machined surfaces provide better conditions for this seal. When the rubber gasket is used as a seal against the casing, special care must be taken to assure that the contact surface is clean and smooth. Clamp-and-gasket connections should be designed so that forces resulting from weight, misalignment, twisting, settlement, and vibration are absorbed by the metal parts, and not by the rubber gaskets.

Materials used in adapters, adapter units, and accessories should be selected carefully for strength and resistance of corrosion. Corrosion potential is high closest to the surface and where there is moisture and air. To use metals of differing "potential" in contact with each other in a corrosive environment is to invite rapid destruction of one of them by electrolytic corrosion. For example, steel clamps would work better with steel casing than most other metals or alloys. Some metals that by themselves resist corrosion (e.g.; bronze, brass, copper, aluminum) may corrode, or cause others to corrode, when placed in contact with a different type of metal. Different metals placed in a corrosive environment should be insulated from each other by rubber, plastic, or other nonconductor. Care should be taken in the selection of welding materials; the welded connection is frequently the point where corrosion begins.

Cast iron is more resistant to corrosion than steel under many conditions of soil and water corrosiveness. However, some grades of cast iron are unable to resist severe stresses from tension, bending, and impact. Metals used in castings subjected to such stresses should be chosen, and the parts designed, to meet these requirements. Breakage

---

5 **Some States prohibit the use of "Dresser type" connections for pitless units.**

can be serious and expensive, especially if pumping equipment, pipes, and accessories fall into the well.

For the same reasons, plastics should be used in adapters and units only where they are not subjected to severe forces of bending, tension, or shear.

Excavation around the well produces unstable soil conditions, and later settlement will occur. Unless at least a portion of the line is flexible, settlement of the discharge line will place a load on the adapter connection that could cause it to break or leak. If for some reason the use of rigid pipe is necessary, the connection should be by means of a "gooseneck", a "swing joint", or other device that will adjust to the settlement without transferring the load to the adapter. The best fill material to use to minimize settlement of the discharge line is fine to medium sand, washed into place. With a correctly placed cement grout seal around the casing and below the point of attachment (see Figure 32), the sand will not find its way into the well. Sand does not shrink or crack in drying, and several feet of it form an efficient barrier against bacteria.

Once a pitless unit has been installed and tested, there is still a risk of accidental damage to the buried connection. There have been numerous cases of breakage by bulldozers and other vehicles. Until all construction and grading around the area are finished, the well should be marked clearly with a post and flag. A 2- by 4-inch board, 3 or 4 feet long, clamped or wired securely to the well casing and bearing a red flag, will do the job.

If the well is in an area where motor vehicles are likely to be driven, the final installation should include protective pipe posts set in concrete. The posts should be just high enough to protect the well, but not so high that they interfere with well servicing.

Inspection and Testing of Pitless Devices

Pitless adapters and units are installed within the upper 10 feet of the well structure, the zone of greatest potential for corrosion and contamination. Procedures for inspecting and testing are therefore important.

The buyer should pick an adapter or unit that not only satisfies health department requirements and the design criteria above, but whose manufacturer will stand behind it.

Hiring a contractor with a reputation for good work is probably the best assurance of getting the job done right. The owner should insist that the contractor guarantee his work for at least 1 year. Some state and local health departments maintain lists of licensed or certified contractors authorized by law to construct wells and install pumping systems.

Field connections on pitless adapters and units can be easily tested with the equipment shown in Figure 32. First, the lower plug is positioned just below the deepest joint to be tested, and then inflated to the required pressure. The sanitary well seal is then positioned in the top of the well and tightened securely to form an airtight seal. This isolated section of the casing or unit is then pressurized through the discharge fitting, or through a fitting in the sanitary well seal. A pressure of 5 to 7 pounds per square inch should be applied and maintained, without the addition of more air, for 1 hour. *Warning: Do not hold face over well seal while pressurized!* While under pressure, all field connections should be tested for leaks with soap foam. Any sign of leakage--either by loss of pressure or by the appearance of bubbles through the soap--calls for repair and retesting.

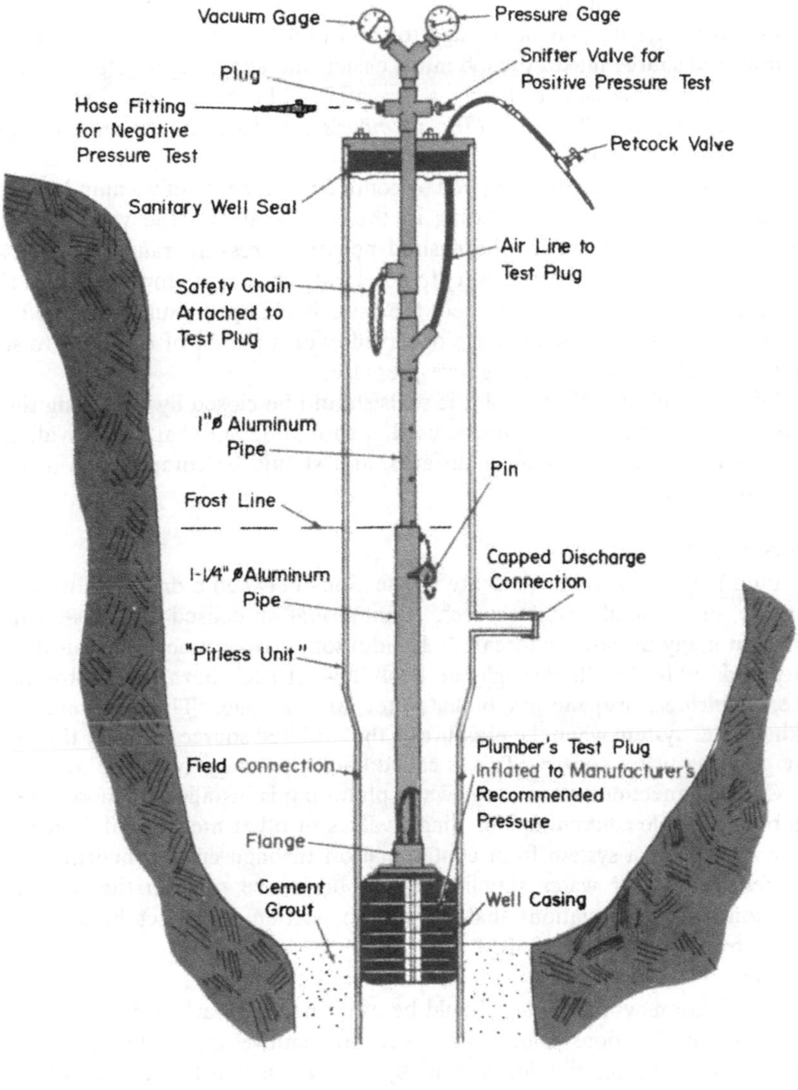

FIGURE 32. - ***Pitless adapter and unit testing equipment.***

Adapters and units that depend on rubber or plastic seals in the field connection should also be tested under "negative" pressure conditions. This can be accomplished by connecting the hose fitting (Figure 32) to a source of vacuum. The negative pressure is read on the vacuum gauge.

Positive pressure may be applied to the isolated section by means of a tire pump, but a powered source makes the job much easier and encourages better testing. If an air compressor is not available or handy, a tire-inflation kit of the kind that uses automobile engine compression will work. *(The plumber's test plug should only be inflated by a hand-operated tire pump).*

Negative pressure can be applied by connecting a length of vacuum hose (heavy wall, small bore) between the hose fitting in the well seal and the vacuum system of an automobile engine. To reach the desired negative pressure range (10 to 14 inches of mercury vacuum), it may be necessary to accelerate the engine for a period of time. Once the desired pressure range is reached, the hose is clamped shut or plugged, the engine disconnected, and the vacuum gauge observed over a period of one hour to see whether there is any detectable loss of negative pressure.

Leaks found in rubber or plastic seals should be closed by tightening the clamps, if possible. If a cement sealant must be used, it should be one that will provide a strong yet flexible bond between the sealing surfaces, and should be compounded to provide long service when buried.

CROSS-CONNECTIONS

Plumbing cross-connections are connections between a drinkable water supply and an unsafe, or polluted, water source[6]. Contamination caused by cross-connections has resulted in many disease "outbreaks." In addition, cross-connections can threaten water quality and public health through the back-flow of such hazardous substances as anti-freeze, herbicides, propane gas, boiler water, and sewage. The contaminant enters the drinking water system when the pressure of the polluted source is higher than the pressure of the drinking water source. This is called back-siphonage or back-flow.

Cross-connections often occur when plumbing is installed by those unaware of the dangers of possible contamination. Single valves or other mechanical devices may not be enough to protect a system from contamination through cross-connections. Therefore, those responsible for water supplies should be aware of both the dangers of cross-connections and of situations that require inspection to detect hazardous conditions resulting from cross-connections.

Detection

Distribution system piping should be examined regularly to identify potential cross connections, and actions should be taken to remove any potential health hazards. Connections to the distribution system by a location which uses hazardous chemicals should be identified and adequate cross connection protection provided. All connections between safe and contaminated water should be identified and removed.

---

6 **U.S. Environmental Protection Agency, *Cross Connection Control Manual*, EPA, Office of Water, Washington D.C. (1989).**

Prevention

There are many types of devices useful for preventing backflow at cross-connections. The device should be chosen according to the degree of hazard posed by the cross-connection. In addition, piping size, location and the need to test the device regularly must be considered.

Air Gap. These are non-mechanical backflow preventors and work well to prevent back-siphonage and back-pressure conditions. Air gaps stop the piping flow with a loss of pressure. Therefore, they are generally used at the ends of the line service. Each air gap requires a reservoir and secondary pumping system.

Barometric Loop. The loop consists of a continuous section of supply piping that abruptly rises to a height of approximately 35 feet and returns back to the original level. It may be used to protect against back-siphonage, but not against back-pressure.

Atmospheric Vacuum Breaker. The breaker is constructed of a polyethylene (plastic) float which travels freely on a shaft and seals in the uppermost position against a silicone rubber disc. Water flow lifts the float and keeps it in the upper sealed position. This device is one of the simplest and least expensive back flow preventors. However, it cannot be used as protection against back-pressure.

Hose Bibb Vacuum Breaker. This device is a vacuum breaker which is usually attached to sill cocks, which serve garden hose type outlets. It consists of a spring-loaded check valve that seals against an air outlet once water pressure is applied. Once the water supply is off, the breaker vents to the air, thus preventing back siphonage.

Pressure Vacuum Breaker. This is another special type of atmospheric vacuum breaker which can be used under constant pressure. It does not protect against back-pressure, and must be installed at least 6 to 12 inches higher than the existing outlet.

Double Check with Intermediate Atmospheric Vent. This unique backflow preventor, which is useful in 1/2-inch and 3/4-inch piping, is used under constant pressure and protects against back-pressure. Construction is basically a double check valve with a vent to air located between two check valves.

Double Check Valve. This is essentially two check valves in one casing with test cocks and two tightly closing gate valves. Double check valves are commonly used to protect against low to medium hazard sources. They protect against both back-siphonage and back-pressure conditions.

Double Check Detector Check. An outgrowth of the double check valve, this device is usually used in fire lines. The device protects the potable water source from fire fighting chemicals, booster pump fire line backpressure, stagnant "black water", and additional "raw" water from outside fire pumper connections. It is constructed with test cocks to insure proper operation of the primary check valves and the by-pass check valve. The valve permits normal usage flows to be metered through a by-pass system with minimal pressure drop. However, in conditions of high flow (such as fires) water passes with minimal restriction through two large spring-loaded check valves.

Residential Dual Check. This device gives reliable and inexpensive protection from back-pressure. It is sized for half, three-quarter and one inch service lines and is installed immediately downstream of the water meter. Plastic check modules with no test cocks or gate valves lower the cost of this device.

Reduced Pressure Principle Backflow Preventor. This is a modified double check valve with an atmospheric vent placed between the two checks. The device provides protection from back-siphonage and back-pressure and can be used under constant pressure in high hazard conditions.

PIPE AND FITTINGS

For reasons of economy and ease of construction, many small public water systems use plastic pipes and fittings. Other types of pipes used are cast iron, asbestos-cement, concrete, galvanized iron, steel and copper. Under certain conditions and in certain areas, it may be necessary to use protective coatings, galvanizing, or have the pipes dipped or wrapped. When corrosive water or soil is encountered, coated copper, brass, wrought iron, plastic or cast iron pipe, although usually more expensive initially, will have a longer, more useful life.

Pipes should be laid as straight as possible in trenches, with air-relief valves or hydrants located at the high points on the line. Failure to provide for the release of accumulated air in a pipeline on hilly ground may greatly reduce the capacity of the line. It is necessary that pipeline trenches be deep enough to prevent freezing in the winter. Pipes placed in trenches at a depth of more than three feet will also help to keep the water in the pipeline cool during the summer months.

Plastic Pipe

Plastic pipe for cold water is simple to install, has a low initial cost, and has good hydraulic properties. When used in a domestic water system, plastic pipe should be certified by an acceptable testing laboratory (such as the National Sanitation Foundation) as being nontoxic and non-taste-producing. It should be protected against crushing and from attack by rodents.

PVC, or polyvinylchloride, pipe is the most widely used plastic pipe. This pipe cannot be thawed electrically, and so creates problems in colder climates where freezing might occur. It is fairly flexible, lightweight, resistant to corrosion and has a long service life. A special coating must be applied to prevent deterioration of the pipe if exposed to sunlight.

Cast Iron Pipe (CIP)

This pipe has high resistance to corrosion and great strength. A thin coating of cement mortar on the inside of the pipe lessens the likelihood of corrosion within the pipe and provides a more friction-free inside surface. Cast iron is not usually available in sizes below 2 inches in diameter, so its use is restricted to larger transmission lines.

Ductile Iron Pipe (DIP)

Ductile iron pipe is similar to cast iron pipe, but is stronger, more flexible ("ductile"), lighter and offers corrosion resistance similar to or better than cast iron. It is normally available in 3-inch and larger sizes. Cement mortar is often used to line the inside of ductile iron pipe to reduce internal corrosion, and a bituminous coating or polyethylene wrap may be applied to reduce external corrosion. When properly coated and lined, ductile iron pipe provides an extremely long service life.

Asbestos Cement Pipe (ACP)

This type of pipe is easy to install and has moderate corrosion resistance. However, due to its non-conductivity, electrical thawing is impossible. In addition, it is rather fragile and must be handled, imbedded and installed carefully. In addition, when "tapping" into

ACP lines, care must be taken to prevent a release of asbestos debris or fibers into the water supply. The risk of releasing asbestos when in contact with corrosive water is also a concern. Due to these factors, this type of pipe is rarely encountered in newer systems.

Reinforced Concrete Pipe

Typically used for larger transmission and distribution mains the concrete pipe is most economical because of its durability, strength, and carrying capacity. However, it is difficult to handle, due to its heavy weight and resistance to tapping.

Steel Pipe

Steel pipe is strong, lightweight, easy to handle and transport, and easily assembled. This pipe is prone to corrosion and needs to be lined and coated to prolong its life.

Copper Pipe

This type of pipe has high resistance to corrosion and is easily installed. Fittings are usually available in the same sizes and materials as piping, but valves are generally cast in bronze or other alloys. In certain soils, the use of dissimilar metals in fittings and pipe may create electrolytic corrosion problems. The use of nonconductive plastic inserts between pipe and fittings or the installation of sacrificial anodes is helpful in minimizing this type of corrosion.

PIPE CAPACITY AND HEAD LOSS

The pipe selected should be large enough to deliver the required peak flow of water without excessive loss of pressure. The normal operating water pressure for household use ranges from 20 to 60 pounds per square inch,[7] or about 45 to 140 feet of head at the fixture.

The capacity of a pipeline is determined by its size, length, and interior surface condition. Assuming that the length of the pipe is fixed, and its interior condition known, the key problem in design of a pipeline is that of determining the right diameter of pipe.

The correct pipe size can be selected with the aid of Figure 33, which gives size as a function of pressure (head) loss, H, length of pipeline, L, and peak discharge, Q. For example, suppose that a home and farm installation is served by a reservoir a minimum distance of 500 feet from the point of use, whose surface elevation is at least 150 feet above the level of domestic service, and in which a minimum service pressure of 30 pounds per square inch is required. It will be necessary first to determine the maximum operating head loss, i.e., the difference in total head and the required pressure head at the service.

$$H = 150 - (2.3 \times 30) = 150 - 69 = 81 \text{ feet}$$

The maximum peak demand which must be delivered by the pipeline is determined to be 30 gallons per minute based on local usage rates.

$$Q = 30 \text{ gallons per minute}$$

---

7 One pound per square inch is the pressure produced by a column of water 2.31 feet high.

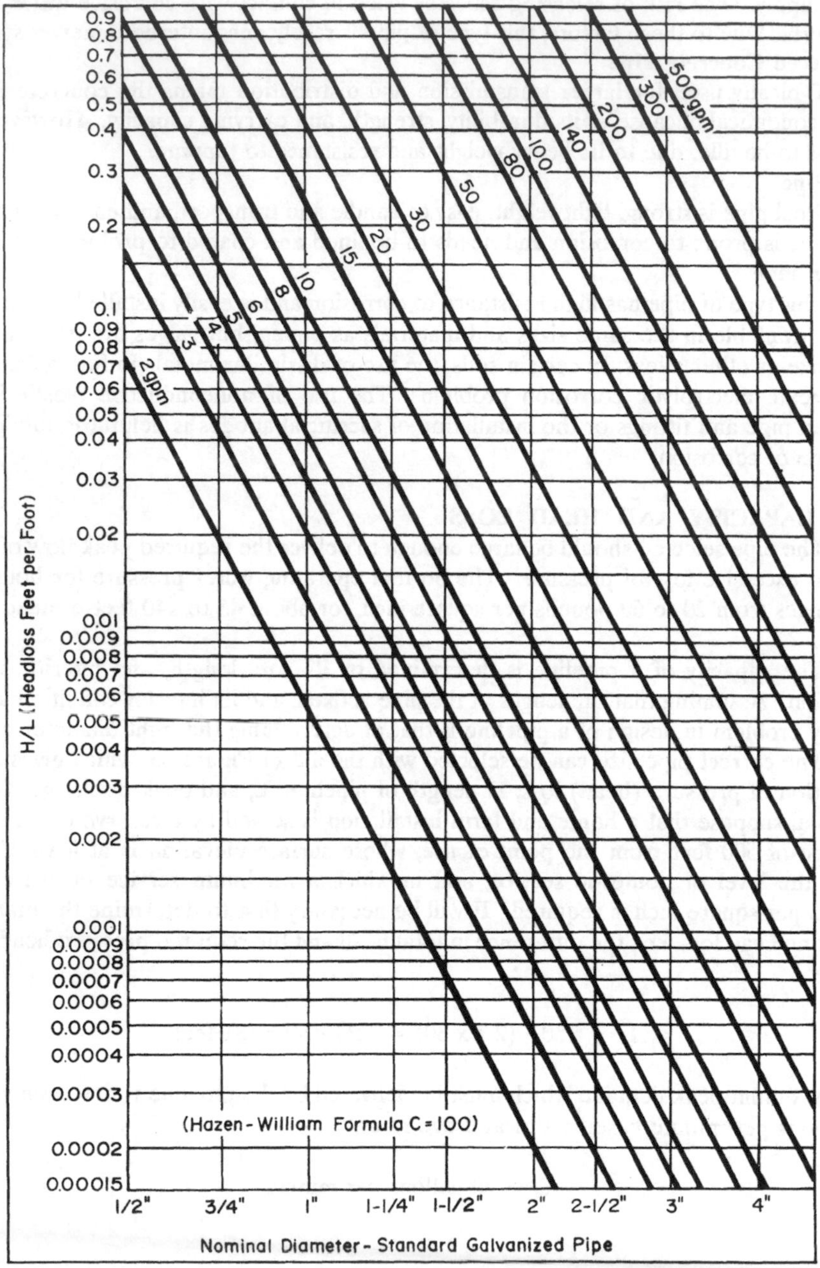

FIGURE 33. - ***Head loss versus pipe size.***

The hydraulic gradient is 0.162 foot per foot.

$$\frac{H}{L} = \frac{81}{500} = 0.162 \text{ foot per foot}$$

Using Figure 33, with the computed values of H/L and Q, one finds that the required standard galvanized pipe size is approximately 1 3/8 inches. Since pipes are available only in standard sizes, standard pipe of 1½ inches in diameter (the next larger size) should be used.

Additional pressure losses may be expected from the use of fittings in the pipeline. These losses may be expressed in terms of the equivalent to the length and size of pipe which would produce the same loss if, instead of adding fittings, we added additional pipe. Table 10 lists some common fitting losses in terms of an equivalent pipe length.

In the example given above the inclusion of two gate valves (open), two standard elbows, and two standard tees (through) would produce a pressure loss equivalent to 15 feet of 1½-inch pipe. From Table 10, one finds that by using 515 feet of 1½-inch pipe instead of the actual length of 500 feet (H/L=0.157), the capacity of the system for the same total head loss is about 38 gallons per minute. This is a satisfactory flow.

TABLE 10. - ***Allowance in equivalent length of pipe for friction loss in valves and treaded fittings***

Diameter of fitting	90 degree ell	45 degree ell	90 degree tee	Straight run	Gate valve	Globe valve	Angle valve
Inches	Feet	Feet	Feet	Feet	Feet	Feet	Feet
3/8	1	0.6	1.5	0.3	0.2	8	4
1/2	2	1.2	3	0.6	0.4	15	8
3/4	2.5	1.5	4	0.8	0.5	20	12
1	3	1.8	5	0.9	0.6	25	15
1-1/4	4	2.4	6	1.2	0.8	35	18
1-1/2	5	3.0	7	1.5	1.0	45	22
2	7	4.0	10	2	1.3	55	28
2-1/2	8	5.0	12	2.5	1.6	65	34
3	10	6.0	15	3	2	80	40
3-1/2	12	7.0	18	3.6	2.4	100	50
4	14	8.0	21	4	2.7	125	55
5	17	10.0	25	5	3.3	140	70
6	20	12.0	30	6	4	165	80

It can be seen from this example that fitting losses are not particularly important for fairly long pipelines, say greater than about 300 feet. For pipelines less than 300 feet, fitting losses are very important and have a direct bearing on pipe selected; therefore, they should be calculated carefully.

Globe valves, which produce large pressure losses, should be avoided in main transmission lines for small water systems.

Interior piping, fittings, and accessories should conform to the minimum plumbing requirements of the *National Plumbing Code*[8] or the applicable plumbing code of the locality.

PROTECTION OF DISTRIBUTION SYSTEMS

The sanitary protection of new or repaired pipelines can be ensured by close attention to certain details of construction. All connections should be made under dry conditions, either in a dry trench or, if a dry trench is not possible, above the ground on a dry surface. Soiled piping should be thoroughly cleaned and disinfected before connections are made. Flush valves or cleanouts should be installed at low points where there is no possibility of flooding.

When not properly designed or installed, frostproof hydrants may allow contamination into the water system. Such hydrants should be provided with good drainage to a free atmosphere outlet where possible. The drainage from the base of the hydrant should not be connected to a seepage pit that is subject to pollution, or to a sewer. The water-supply inlet to water tanks used for livestock, laundry tubs, and other similar installations should be placed with an air gap (twice pipe diameter) above the flooding level of the fixtures to prevent back siphonage. There should be no cross-connection, auxiliary intake, bypass, or other arrangements that would allow polluted water, or water of questionable quality, to be discharged or drawn into the domestic water supply system.

Before a distribution system is used, it should be completely flushed and disinfected.

DISINFECTION OF DISTRIBUTION SYSTEMS

General

The system's distribution system should be disinfected if untreated or polluted water has been in the pipe, upon completion and before operating the new system to insure water of satisfactory quality, and after maintenance and repair.

Procedure

The entire system, including tanks and standpipes, should be thoroughly flushed with water to remove any sediment that may have collected during construction. After flushing, the system should be filled with a disinfecting solution of calcium hypochlorite. This solution is prepared by adding 1.2 pounds of high-test 70 percent calcium hypochlorite to each 1,000 gallons of water, or by adding 2 gallons of ordinary household liquid bleach to each 1,000 gallons of water. A mixture of this kind provides a solution of at least 10 mg/L of available chlorine.

The disinfectant should be left in the system, tank, or standpipe, for at least 24 hours, examined for residual chlorine, and drained out. If no residual chlorine is found, the process should be repeated. Next, the system should be flushed with treated water and tested for coliform. After coliform test results are satisfactory, the system, tank or stand pipe can be placed into operation.

---

[8] **Obtainable at the American Society of Mechanical Engineers, United Engineering Center, 345 East 47th St., New York, N.Y. 10017.**

DETERMINATION OF STORAGE VOLUME

Three types of storage facilities are commonly employed for individual water supply systems. These are pressure tanks, elevated storage tanks, and ground-level reservoirs and cisterns.

When ground water sources with sufficient capacity and quality are used, only a small artificial storage facility may be needed, since the water-bearing formation itself is a natural storage area. However, if the well is not able to meet peak water demand or treatment is required, additional storage volume will be needed.

If water demand doesn't change, and treatment or well capacities can be increased, the amount of storage required will decrease. Therefore, there is a balance between providing a larger treatment process or well capacity and providing additional storage.

Pressure Tanks

Pressure in a distribution system with a pneumatic tank is maintained by pumping water into the tank. This pumping action compresses a volume of entrapped air. The air pressure is equal to the water pressure in the tank and can be kept between desired limits by using pressure switches. These switches stop the pump at the maximum setting and start it at the minimum setting. The capacity of the pressure tank is usually small when compared to the total daily water consumption. Only 10 to 40 percent of a pressure tank volume is usable storage. For this reason, pressure tanks are only designed for peak demands. The maximum steady demand the system can deliver is equal to the pump capacity.

The usable storage of a pressure tank can be increased by "supercharging" it with air when it is installed, or by recharging at the factory. Recharging can only be done in tanks in which the water space and air space is completely separated by a diaphragm or bladder. Consult your dealer for design details and characteristics.

Use the figures in Tables 11 and 12 for the selection of pumps and pressure tanks for various size homes. The pump capacity from Table 11 can be used to find the right tank size for the type (precharged, supercharged, or plain) and pressure range needed. These tabulated values are recommended by the Water System Council[9].

When a pressure tank is part of the distribution system, there will be no problem with "water hammer". Otherwise, it may be necessary to provide an air chamber on the discharge line from the well, located near the pump, to minimize water hammer.

TABLE 11. - *Seven-minute peak demand period usage.*

Number of baths in home:	1	1½	2-2½	3-4
Normal 7-minute peak demand (gal.)	45	75	98	122
Minimum size pump to meet demand without using storage	7 gpm	10 gpm	14 gpm	17 gpm

NOTE: Values given are average and do not include higher or lower extremes.

9 Water Systems Council, 221 North LaSalle Street, Chicago, IL 60601

TABLE 12. - *Tank selection chart - gallons (Based on present industry practice)*

Pump capacity		Minimum Draw Down	Switch setting pounds per square inch								
			20-40			30-50			40-60		
gpm	gpm	(gal)	*A	*B	*C	*A	*B	*C	*A	*B	*C
240	4	4	15	15	20	15	20	30	20	20	40
300	5	5	15	20	30	20	25	40	25	25	50
360	6	6	20	20	35	25	25	45	30	30	55
420	7	7	20	25	40	25	30	55	30	40	75
480	8	8	25	30	40	30	35	65	35	45	85
540	9	9	30	30	50	35	40	70	40	50	95
600	10	10	30	35	55	40	45	80	45	55	105
660	11	12	35	40	60	45	50	95	55	65	125
720	12	13	40	45	70	50	60	105	60	70	135
780	13	15	45	50	80	60	65	120	70	80	155
840	14	17	55	60	90	65	75	135	75	90	175
900	15	19	60	65	100	75	80	150	85	105	195
960	16	20	65	70	110	75	90	160	95	115	205
1020	17	23	70	80	120	90	100	185	105	125	240
1080	18	25	80	85	135	95	110	200	115	140	260
1140	19	27	85	95	150	105	120	215	125	150	280
1200	20	30	95	105	160	115	130	240	140	165	310

*A - Precharged bladder or diaphragm tank.
*B - Supercharged, floating water tank.
*C - Plain steel tank.

Elevated Storage

Elevated tanks should have a capacity of at least two days of average consumption. Larger storage volume may be necessary to meet special demands, such as firefighting or equipment cleanup.

Ground-Level Reservoirs and Cisterns

Reservoirs that receive surface runoff should generally be large enough to supply the average daily demand over a dry period of the maximum length anticipated. Cisterns are usually designed with enough capacity to provide water during periods of less than one year.

PROTECTION OF STORAGE FACILITIES

Suitable storage facilities for relatively small systems may be made of concrete, steel, brick, and sometimes wood (above the land surface). Such storage facilities should receive the same care as system installations in the selection of a suitable location and protection from contamination. Waterproofing the interior of storage units with asphalt or tar is not recommended because of the unpleasant taste imparted to the water, and the possibility of chemical reactions with materials used for treatment. Specifications for painting water tanks are available from the American Water Works Association.[10] Appropriate federal, state, or local health agencies should be consulted about acceptable paint coatings for interior tank use.

All storage tanks for domestic water supply should be completely covered and so constructed to prevent pollution by outside water or other foreign matter. Figures 34 and 35 show some details for manhole covers and piping connections to prevent pollution by surface drainage. Concrete and brick tanks should be made watertight with a lining of rich cement mortar. Wood tanks are usually constructed of redwood or cypress and, while filled, will remain watertight. All tanks require good screening of any openings to protect against the entrance of small animals, mosquitoes, flies, and other small insects.

Tanks containing water to be used for livestock should be partially covered and constructed so that cattle will not enter the tank. The area around the tank should be sloped to drain away from the tank.

Figure 34 shows a typical concrete reservoir with screened inlet and outlet pipes. This figure also illustrates the sanitary manhole cover. The rim should be elevated at least four inches above the ground with the cover extending two inches beyond the edge of the rim. This type of manhole frame and cover should be designed so that it may be locked to prevent access by unauthorized persons.

An emergency water supply, that has been polluted at its source or in transit, should not be added to storage tanks, cisterns, or pipelines used for drinking water.

Disinfection of storage facilities after construction or repair should be carried out in accordance with the recommendations given under "Disinfection of Water Distribution System" in this part of the manual.

It can be seen from this example that fitting losses are not particularly important for fairly long pipelines, say greater than about 300 feet. For pipelines less than 300 feet, fitting losses are very important and have a direct bearing on pipe selected; therefore, they should be calculated carefully.

Globe valves, which produce large pressure losses, should be avoided in main transmission lines for small water systems.

Interior piping, fittings, and accessories should conform to the minimum plumbing requirements of the *National Plumbing Code*[11] or the applicable plumbing code of the locality.

---

10 American Water Works Association, 6666 West Quincy Avenue, Denver, Colorado 80235.

11 Obtainable at the American Society of Mechanical Engineers, United Engineering Center, 345 East 47th St., New York, N.Y. 10017.

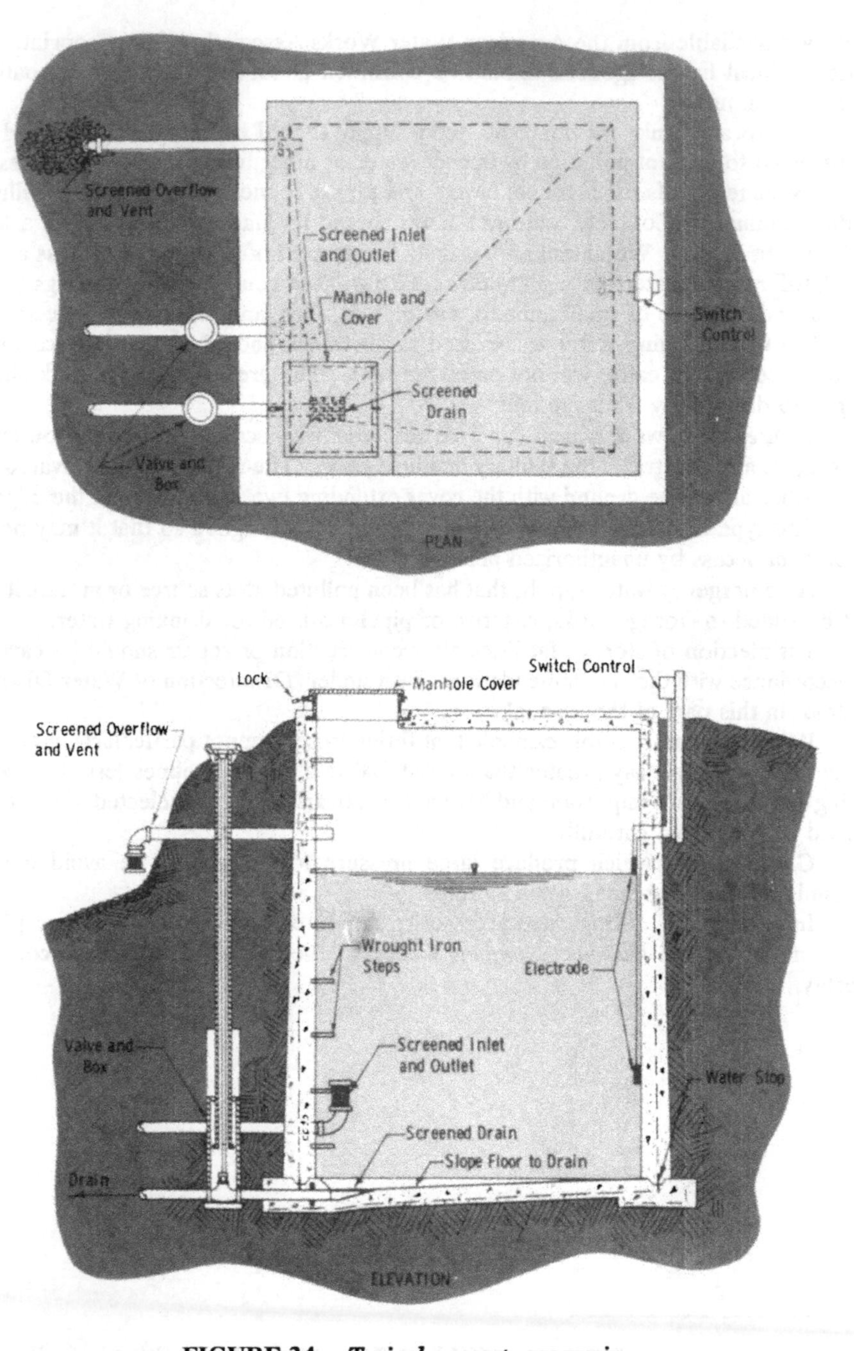

FIGURE 34. - *Typical concrete reservoir.*

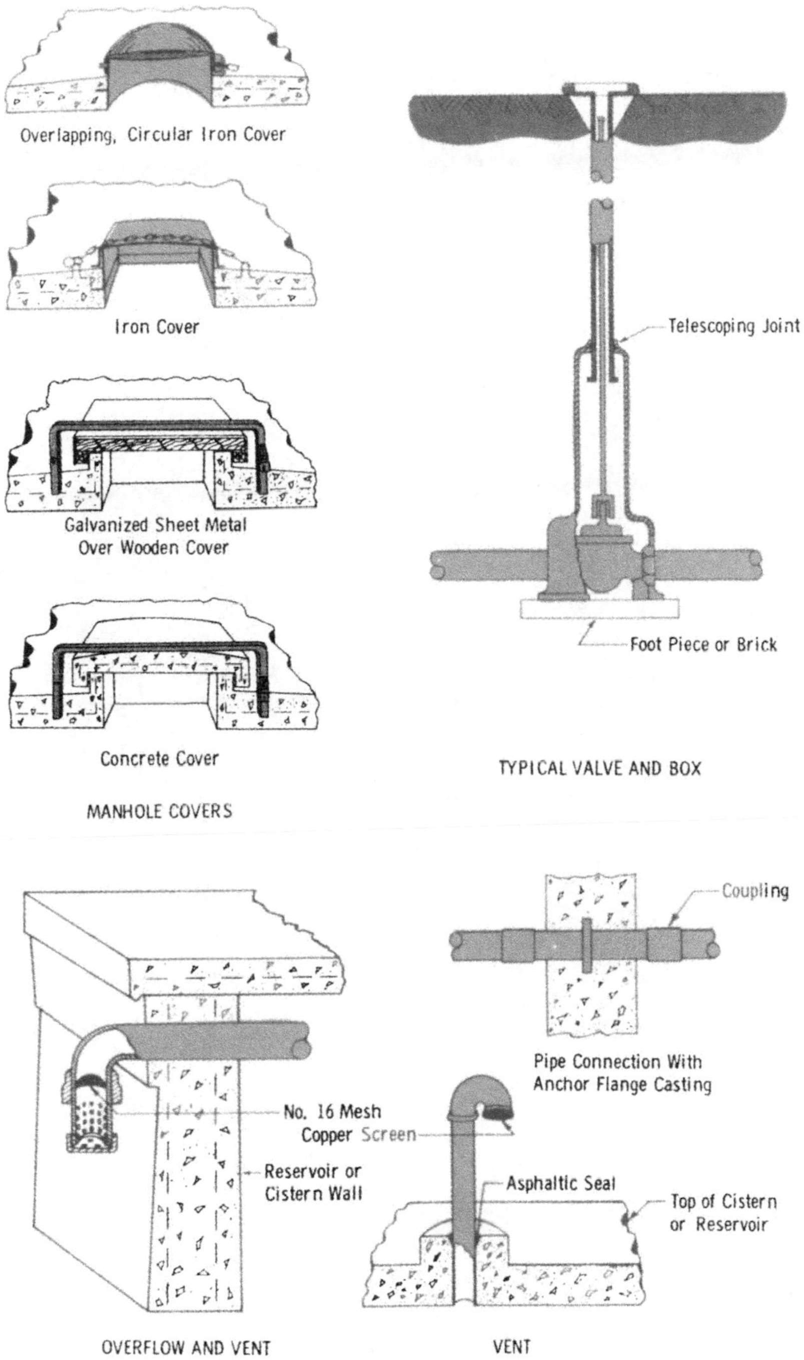

FIGURE 35. - *Typical valve and box, manhole covers, and piping installations.*

Part VI

Information, Assistance & Community Support

INTRODUCTION

Depending on specific needs, a small community water supply system may take various steps to improve water quality and/or supply. These steps may include selecting new equipment, or pursuing ways to improve operations and maintenance of the water supply system. Other means of improving the water system may include:

- Building a better understanding of regulations affecting public and private water systems.
- Seeking financial and technical support.
- Locating trained personnel to install, operate and/or maintain the water system.
- Improving community awareness of water supply issues.
- Creating a climate for active community participation.
- Finding organizations to assist in water supply planning.

Parts I through V of this manual provide information regarding the selection of a suitable water source and the treatment techniques necessary to ensure delivery of safe drinking water. For this information to be best used, it may also be necessary for a small water system to obtain current information and locate other means of technical and financial support. The purpose of this section is to provide direction in identifying such resources available to communities.

INFORMATION RESOURCES

The first and most important tool is *information*. A small public water system must gain information and an understanding of local and federal requirements that apply to water supply. The county or other local health authority may be the first contact for this. State drinking water agencies are also excellent information sources (Appendix F). EPA drinking water offices (Appendix G) are also available to help. EPA also operates a Safe Drinking Water Hotline (1-800-426-4791), so that citizens may obtain current information regarding EPA drinking water programs.

Other organizations provide information or support for small water systems. These include the National Rural Water Association (NRWA), American Water Works Association (AWWA), National Water Supply Improvement Association (NWSIA), and the Rural Community Assistance Program (RCAP). The address and phone number of these and other organizations can be found in Appendix H. Other services provided by these organizations are described in the Technical and Financial Assistance sections of this part.

Another source of valuable information includes the accumulated knowledge and experience of other similar-sized water suppliers. Systems in the same geographical area face many of the same challenges, and their problem solving experience can be shared. A single contact of this type can often provide valuable information and guidance.

Information is also available in many references, which provide detailed explanations of the various assistance options, source protection and development methods, and treatment techniques described in this manual. Useful references are included in the Bibliography.

Public Education

An informed public can be a valuable ally in the effort to provide a safe and reliable supply of drinking water. A community needs to be aware of the Safe Drinking Water Act (SDWA), its purpose, and the impact these regulations may have on their water system--especially those regulated under SDWA as public water systems. Public education is a responsibility shared by many, including the various levels of government, water industry organizations, and the water system owners.

Information on water supply is provided in several forms:

- Distribution of informative pamphlets.
- Monthly newsletters or public service press releases to the news media.
- Events such as seminars, water works conventions, etc.
- Public appearances for TV, radio and local clubs.

Promoting or attending seminars on drinking water quality, health effects and water source protection will increase public knowledge and interest in water supply. These can be included with special events such as water awareness week, county health programs, school tours or other local activities. This can be an effective means of building a community's interest in drinking water issues.

Speaking at local community functions about water supply offers an opportunity for an owner or operator to increase public knowledge of drinking water. This includes informing the public of new water regulations and what is being done locally to improve public water supply.

COMMUNITY INVOLVEMENT

A second powerful tool is the *organization* of information and resources. The organization may be a person or group of persons in the community who focuses on water supply, studies and understands the issues, participates in decision making and communicates with the rest of the community. A water supply committee may best serve as the organizational unit to perform these functions, but in some cases, less formal community participation may be as effective. An organized, informed community can be actively involved in the decision making and long-term planning goals, to help create a safer and more reliable drinking water supply. Supported by the community and other organizations (see Figure 36), a water supply committee can do the following:

- Assist in assuring protection of water source.
- Gather important information about available water resources, regulations, health issues, etc.
- Set priorities for water system.
- Plan and schedule water supply improvements.
- Select contractors, equipment suppliers, engineers, and staff as required.
- Review operations and maintenance procedures periodically.
- Keep abreast of system finances and available sources of funding/loans.
- Seek additional training as needed.
- Report regularly to the community.

TECHNICAL ASSISTANCE AND TRAINING

Many sources of assistance for small water systems can be found on Federal, State and local levels and from groups such as the American Water Works Association. These sources can lend technical assistance in the form of technical training, hands-on operational assistance, design support, and regulatory guidance.

Technical Support

Many states have programs specifically designed to provide technical support to small community water supply systems. Operator training and certification is an established program in more than 40 states. State programs offer a wide range of training from beginner level correspondence courses to advanced, comprehensive water treatment courses. In most states, these programs are organized by the state drinking water agencies (see Appendix F).

Large water utilities can sometimes help small systems with technical assistance. In some areas "circuit riders" are available to help with the operation and maintenance of systems that cannot hire a local person to take on those responsibilities. Circuit riders are licensed operators who provide on-site technical assistance in the operation of a number of small water systems within a defined service area. These programs are provided by many state regulatory agencies, the National Rural Water Association (NRWA), larger utilities and individual operators. A circuit rider assistance program can also be written into service contracts with utilities, consulting firms, or equipment vendors.

The Rural Community Assistance Program (RCAP) can also help rural community officials build their skills in solving local water problems. It provides on-site technical assistance, training and publications, and works to improve federal and state government responsiveness to the needs of rural communities.

In addition, local engineering consultants and equipment manufacturers representatives are available to advise on matters such as technology, installation, operations, etc.

Operations and Maintenance

Operations and maintenance (O&M) is a central concern of a water supply, no matter how simple or complex the system. Operations refers to those activities performed

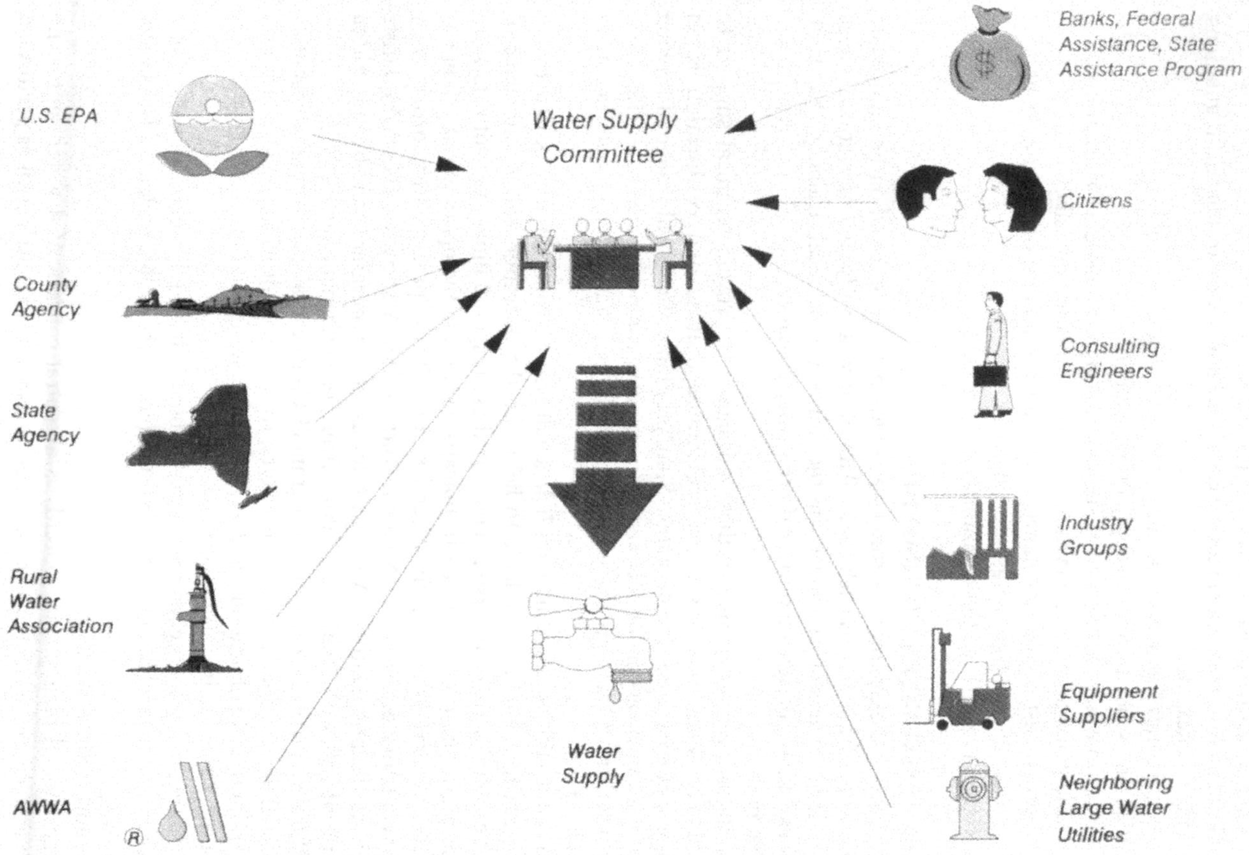

FIGURE 36. - ***Sources of information and assistance.***

on a regular basis to ensure that the water system reliably delivers an uninterrupted supply of potable water. Operating treatment equipment, adding chemicals, ordering treatment supplies and spare parts, performing water quality analysis, and keeping records are typical operational duties. Maintenance refers to work carried out to ensure that equipment and related materials are able to function, such as periodic cleaning, lubrication or repair of equipment.

A process control program involving laboratory testing and adjustment of chemical feed equipment to changes in water quality is a key element in maintaining system performance. Through training and experience, operators become skilled in evaluating changes in raw water quality and making required chemical feed changes to maintain a consistent high-quality finished water.

New procedures, requirements, and occasional staff turnover require an on-going training program for operators and managers of small water systems. This is an essential and continuous process. As mentioned above, state agencies and water industry organizations sometimes offer training suited for small water systems personnel. O&M costs and the cost of replacing major equipment items must be included in the water rate. Loans obtained for water systems may in some cases include provisions requiring revenucs to cover long-term O&M costs, or a service agreement which provides the O&M needs of a water system. Proper O&M guarantees better long-term performance and more cost-effective operation of a water supply.

Several agencies provide on-site assistance for selection of a general consultant, laboratory services, design review, utility management, and rate setting. There are also operation and maintenance manuals available for the small water system, some of which are listed in the Bibliography of this manual.

FINANCIAL ASSISTANCE

Many small water supply systems which only have a few customers over which to spread their costs may require financial assistance in order to make improvements to their system. Many federally-funded and state-sponsored financial assistance programs are outlined below.

Farmers Home Administration (FmHA)

The FmHA of the U.S. Department of Agriculture has several programs that provide grants and loans. These loans and grants are the most important outside source of money available to many small water utilities throughout the United States. The FmHA manages several programs that aid small publicly-owned water utilities. Preference is given to projects serving low income communities and mergers of small facilities.

Borrowers of FmHA funds must be unable to secure loans from other sources at reasonable rates or terms. They must be financially sound and able to borrow and repay loans, to pledge security for loans, and to operate the facilities. Grants may be used to cover up to 75 percent of facility costs and can be used along with other financial assistance. The FmHA will assist those seeking funds in assembling information on the engineering feasibility, economic soundness, financing and other aspects of the project.

The FmHA programs provide incentives to prevent building poorly planned water systems, and to encourage regional systems. FmHA also urges small water systems to combine, and generally makes funds available contingent upon cooperative agreements.

Appendix I contains a listing of FmHA state offices.

Community Planning and Development

The Community Planning and Development Agency of the U.S. Department of Housing and Urban Development (HUD) has Community Development Block Grants (CDBG) available to cities smaller than 50,000 people. Priority is given to cities with low and moderate income. Water and sewer projects are eligible. The number of very small cities (under 1,000 persons) receiving grants has been increasing. The CDBG program has been a good source of construction funding for small water systems in rural areas and on urban fringes. All but three states (New York, Maryland, Hawaii) operate the program on a state basis.

In many states, the Department of Community Affairs administers the program. In others, the Department of Economic and Community Development, state planning agencies, the governor's office and economic or industrial development agencies may operate the program. Generally, these state agencies offer financial aid and assistance in applying for grants.

Economic Development Administration (EDA)

The EDA of the U.S. Department of Commerce provides grants for public works and development facilities in economically distressed areas and areas in which improved facilities will encourage the industrial or commercial climate of the area. The EDA program is considered less useful than some of the other federal programs for smaller water systems, but in many cases it has been combined with a FmHA loan program to provide a financially possible project.

U.S. Small Business Administration (SBA)

The SBA assists small businesses with two types of loans. The first type is a bank loan, guaranteed by SBA, with a maximum loan amount of $500,000. The second type is a direct loan made by SBA to a small business that cannot obtain a SBA guaranteed loan from a private source. The largest amount under this direct loan is $150,000. The advantage of the SBA loan program is that it is available to a privately owned utility. Even though the funds are limited, it has been used for improvements, extensions and upgrading of facilities.

Revolving Funds

Several states have revolving funds for providing loans and/or grants to water systems needing capital improvements. To whom these funds are given is normally decided based on need, with priority given to systems which require improvements to comply with water quality regulations. While most funding is in the form of low-interest loans, grants may be available for systems which serve low-to-moderate income levels and have limited financial resources to repay a loan.

The availability and types of grants and/or loans vary among states. Your state drinking water agency can provide information on what is available in your state.

State Loan and Grant Programs

State agencies also may provide funds to small systems to expand or improve facilities in order to comply with the Safe Drinking Water Act or to provide an adequate supply of quality drinking water. However, most states do not provide funding to privately owned facilities under the grant or loan programs.

Industrial Development Bonds

An often ignored source of low cost financing available to small privately owned water utilities is the industrial bonding authority of each state, county or municipality. Pursuant to Section 103(b) of the Internal Revenue Code, each state may issue tax-exempt bonds to be used for private business. The code specifically allows for furnishing water for any purpose so long as the water is made available to all members of the public and the facilities are either operated by a governmental agency or the rates are regulated by a governmental agency. Contact the appropriate state agency for more information on how this program is applied in your state.

Revenue Bonds

A common method of financing improvements to water systems today is revenue bonds. Although not true "assistance", they are a useful means of financing. These bonds are issued and backed by the revenues expected to be received from rates charged for water consumed by the customers. The rates set must be enough to cover the operation and maintenance of the system, provide funds to pay principal and interest to retire the bonds, and meet any other rules as set forth in the bond covenant.

General Obligation Bonds

Another means of financing is through the sale of general obligation bonds, payable from property taxes. These are relatively simple bonds based primarily on the ability of a governmental body to incur debt that does not exceed a fixed percent (usually 10 percent) of the assessed value of property within its control. A second factor involved in issuing these bonds is the requirement in some states that authorization must be obtained by referendum (direct vote).

WATER RATES

The life blood of any utility is an adequate and fair rate structure to provide money for good management, operation and maintenance and amortize (slowly put money aside to repay) any outstanding loans. However, in many of the small and very small water supply systems there is a lack of earnings to accomplish these utility responsibilities, resulting in poor service and poor water quality. Water rates must be set to cover all operational and capital costs in order for a system to remain financially stable. Routine review of the water system's financial health can provide an early warning to prevent financial shortfalls from occurring. A more detailed source of information for determining proper water rates and management practices are available from EPA's booklet *Self-Assessment for Small Publicly-Owned Water Systems*. Additional booklets are available for mobile home parks, privately owned systems, and homeowner associations.

Operating Costs

Operating costs are the costs related to providing and maintaining the water supply. Examples include chemicals and electricity for treatment, maintenance parts, tools, routine replacement of equipment, and payments on loans. One way to be sure that the funds collected to operate a water system are enough to cover operational costs is to calculate the "operating ratio."

The operating ratio is calculated by dividing the total revenues by the total operating costs. Although this is a simple test, it requires a thorough review of accounting records to find the numbers. The following table can be used to figure the operating ratio for a small water system:

OPERATING RATIO WORKSHEET
(Year to Date)

TOTAL REVENUE	
User Service Charges	$ ______
Hook Up/Other User Fees	______
Taxes/Assessments	______
Interest Earnings	______
Other Revenue	______
Total Revenue	$ ______
TOTAL OPERATING EXPENSES	
Administration	$ ______
Wages	______
Benefits	______
Electricity	______
Chemicals	______
Fuel & Utilities	______
Parts	______
Equipment Replacement Fund	______
Principal & Interest Payments	______
Operator Training	______
Contractor Fees	______
Other	______
Total Operating Expenses	$ ______
OPERATING RATIO	
Total Revenue	$ ______
divided by	
Total Operating Expenses	$ ______
equals	
Operating Ratio	______

It is best to calculate the operating ratio on a year-to-year basis since revenues and expenses usually vary from month to month. For a financially healthy water system, an operating ratio of 1.00 is a minimum. Special attention should be given to the trend in the operating ratio. A monthly comparison to previous years during the same month can provide an early warning of trouble so that shortages in funds can be avoided. A stable or upward trend indicates a proper financial balance and that the water rates are properly set.

Capital Costs

Examples of capital costs include those that may be necessary to provide new or improved major equipment or facilities need to meet the requirements of the 1986 SDWA Amendments, to replace aged system parts, or to meet growing demands. Main extensions to serve new customers are usually charged to the developer or those customers. To cover the cost of installing new connections, some systems charge connection fees for materials and labor for the main connection, service line, and meter installation. Connection fees may also include a portion to cover or share the cost of the basic system.

Impacts on water rates from large capital programs or unexpected emergencies can be lessened by careful, long term planning. Many water systems wait too long before increasing user service charges because they fear adverse customer reaction. A good public education program explaining all costs and benefits, together with citizen involvement in water supply, can help to prevent any adverse reaction. Planning for capital improvements, establishing an emergency reserve fund, and increasing rates slowly to meet the financial obligations for improvements can ease the public reaction, and show good management practices.

PERSONAL COMPUTERS

Personal computers are gaining wider use for maintaining operational and financial records for small water systems. Uses include maintaining water pumpage and consumption records, producing reports for regulatory agencies, processing bills, maintaining preventive maintenance program schedules, and calculating revenue needs to cover debt service. Accounting records can also be used for determining returns on equity investment for building reserves or returned earnings, and paying dividends when appropriate.

Special software packages are available through AWWA and other water industry information sources which are fairly simple to use for many of the uses listed and can save time and money. The record maintained by a personal computer is very useful for preparing reports, evaluating system performance and long-term planning.

INSTITUTIONAL ALTERNATIVES

A small, self-contained water system may not have the resources needed to ensure a reliable supply of high quality drinking water in a cost effective manner. As institutional alternatives to independent small water systems, the following organizational structures may be considered:

- Satellites
- Water districts
- Regionalization
- County utilities
- State utilities
- Contract services

An institutional alternative is an organizational structure or procedure that better provides the resources needed to help small water systems provide reliable water service at the lowest possible cost to the consumer. Such resources may include the following:

- Financial
- Legal
- Planning
- Engineering
- Accounting and Collection Services
- Laboratory Support
- Licensed Operators
- Training
- Purchasing
- Supply of Spare Parts
- Consultation on Design
- Treatment and Water Quality Analysis
- Regulatory Liaison
- Leak Detection
- Corrosion Control
- Meter Repair Program
- Emergency Assistance

Satellite of Larger Utilities

A satellite operation refers to the process by which a larger or central water utility assists a small system by one or more of the following:

- Providing technical, operational, or managerial assistance on a contract basis.
- Providing wholesale treated water with or without additional services.
- Assuming ownership, operation, and maintenance responsibility when the small system has a separate distribution system.

A small system is not considered a satellite when it is physically connected to and owned by the larger utility.

The success of a satellite operation is greatly dependent on the economic condition and resources available in the large utility and on the distance, present types of ownership, available finances, political structure and local statutory requirements.

Satellite arrangements may reflect the various ownership types of the larger utilities:

- Investor Owned Systems
- Municipal Systems
- Public Service or Utility Districts
- County Utilities
- State Utilities
- Non-Profit Corporations
- Cooperative Associations

Utility satellites can allow a small water system to benefit from the financial strength of the parent utility, to obtain necessary resources, to improve water quality management and can help secure public funds when the parent utility is publicly owned.

Water Districts

Water districts typically serve water to rural areas and the fringes of urban areas. Water districts are formed by county or local officials (usually upon the petition of citizens) under state laws, to provide one or more water systems in a selected geographical or franchise area. They are sometimes called public service districts, water authorities, or sanitary districts, depending on the state's laws.

Forming a water district is normally a fairly easy task for county commissioners, as it can usually be done by their own motion as well as upon request from consumers. Usually a signed petition has to be presented to force a referendum, although some states require a public referendum prior to setting up a district. Water districts are attractive to county officials as a means of providing water service in rural areas and on the fringes of municipalities having their own systems. They also have proven to be helpful to private utilities that desire to expand but do not have capital. Because public water districts can obtain public funding, a distribution system is often set up in which the larger, private utility then sells water and other services. Water districts, along with property owner associations, have been quite successful in obtaining funds from the Farmers Home Administration (FmHA) which is discussed in the "Financial Assistance" section.

Forming a water district allows the system to be eligible for public grants and loans, issue tax free securities, and help takeovers of independent, small water systems which desire to join the public-owned water system. It may also contract services with publicly owned non-community systems and small privately owned systems.

Regionalization

New and existing small public water supply systems may better serve their customers when merged into a regional system. A regional system is similar to a water district, but is not owned or managed by a government group. A regional system can allow its water systems to take advantage of the economies of scale and good management that can be achieved through larger organizations. Where it is not economically practical to physically interconnect, the water systems can join together to hire a circuit rider to help operate their water treatment plants, or to contract for joint managerial services.

In general, small water systems have very poor economies of scale that can be partly overcome by regionalization. Sharing management and operations personnel, information and equipment resources can allow small water systems to be more able to meet the customers needs.

County Utilities

County utilities are those that are owned and operated by the county (or township) Commissioners or by county Public Works Departments.

Putting a county utility into practice depends on the initiative of the County as well as the willingness of the various boards of health and local authorities to present complaints to the state. Such a complaint may start the formation of a county utility to correct unsafe water supply conditions and protect public health and welfare.

Although they require passage of an enabling law before being put into use, county utilities can provide central management, eligibility for public grants and loans, and allow troubled systems to be taken over by a more capable management organization. County utilities are easy to start and difficult to break up.

State Utilities

State utilities are those that are owned and operated by an agency of state government or are state agents which operate and maintain water utilities for the state on a contractual basis. State utilities include hospitals, camps, colleges, state parks or similar facilities. By providing an umbrella type organization that offers the required resources they allow state government to have a major impact on small water systems. State utilities provide savings through centralized purchasing, management, consultation, planning and technical assistance. This option can also provide coordination with regulatory agencies, sharing of costs of major equipment, and a network of skilled operators.

Contract Services

Contract service is one of the most promising options for solving small water system problems and is used more often as the SDWA provisions go into effect and consumer demand for a quality of service increases.

Contract services may include all or part of the services needed by a small system to allow it to provide good quality and quantity of water. These services could include any one or combination of services such as financial, legal, planning, engineering, accounting and collection, laboratory support, operations assistance, maintenance programs, training, purchasing, supplying spare parts, consulting services, treatment and water quality assistance, regulatory liaison, leak detection, corrosion control, meter repair, emergency assistance and others.

The contract services can provide small water systems with access to needed resources and help them improve water quality management through an easily used option, however, not all services needed may be available through this alternative. Professional organizations such as AWWA and NRWA, the local health departments or State agencies may supply information on firms that provide contract services.

BIBLIOGRAPHY

LIST OF REFERENCES ON SMALL PUBLIC WATER SUPPLY SYSTEMS

American Concrete Pressure Pipe Association, *Basic Water Works Management*, Arlington, VA (1972).

American Public Health Association, American Water Works Association, and Water Pollution Control Federation, *Standard Methods for the Examination of Water and Waste Water*, 17th edition, American Public Health Association, Washington, D.C. (1989).

American Water Works Association, *Water Utility Management*, American Water Works Association, Denver, CO (1980).

American Water Works Association, AWWA Seminar Proceedings, *Small Water System Problems*, American Water Works Association, Denver, CO (1982).

American Water Works Association Research Foundation, *Institutional Alternatives for Small Water Systems*, Denver, CO (1986).

American Water Works Association, AWWA Seminar Proceedings, *Membrane Processes: Principles and Practices*, Denver, CO (1988).

American Water Works Association, *New Dimensions in Safe Drinking Water 2nd edition*, American Water Works Association, Denver, CO (1988).

American Water Works Association, *Water Quality and Treatment*, 4th edition, McGraw-Hill, New York, NY (1990).

American Water Works Association and American Society of Civil Engineers, *Water Treatment Plant Design*, 2nd edition, McGraw-Hill, New York, NY (1990).

Anderson, Keith E., *Water Well Handbook*, Missouri Water Well and Pump Contractors Association, Rolla, MO (1971).

Baker, R.J., Carroll, L.J., and Laubusch, E.J., *Water Chlorination Handbook*, American Water Works Association, New York, NY (1972).

Campbell, Stu, *The Home Water Supply*, Garden Way Publishing, Pownal, VT (1983).

Capitol Controls Co., "Chlorination Guide." Capitol Controls Co., Colmar, PA (undated).

Chang, S.L., "Iodination of Water," *Boletin de la Oficina Sanitaria Panamericana*, Vol. 59, Pages 317-331 (1966).

Cotruvo, J. A. , Perler, A. H., and Bathija, B. L., "Drinking Water Analysis Regulatory Report", Environmental Lab, Pages 22-27 (1989).

DeMers, L. D., Hanson, B. D., Glick, F. L., and Schultz, R. M., "Design Considerations for Package Water Treatment Systems", Journal of The American Water Works Association, Vol. 80, No. 8, Pages 58-61 (1988).

Ficlk, K. J., "New Ozone System Designed Specifically for Small Systems", Waterworld News, Vol. 6, No. 2, Pages 18-19 (1990).

French, W., and Ytell, E., "Rural Community Assistance Program", Journal of the American Water Works Association, Vol. 80, No. 9, Pages 48-51 (1988).

Gollnitz, William D., "Source Protection and The Small Utility", Journal of The American Water Works Association, Vol. 80, No. 8, Pages 52-57 (1988).

Gumerman, R. C., Burris, B. E., and Hanson, S. P., *Small Water System Treatment Cost*, Noyes Data Corporation, Park Ridge, NJ (1986).

Hill, R.D., and Schwab, G.O., "Pressurized Filter for Pond Water Treatment." *Transactions of the ASAE*, Vol. 7, No.4, Pages 370-374, 379, American Society of Agricultural Engineers, St. Joseph, MI (1964).

Hooker, Dan, "How to Protect the Submersible Pump from Lightning Surge Damage," Bulletin DPED-27, General Electric Co., Pittsfield, MA (1969).

Kroehler, C. J., "Underground Injection Control in Virginia", Virginia Water Resources Research Center, Virginia Polytechnic Institute and State University, Blacksburg, VA (1989).

Kroehler, C.J., *What Do The Standards Mean?*, Virginia Water Resources Research Center, Virginia Polytechnic Institute and State University, Blacksburg, VA (1990).

Long, B. W., and Stukenberg, J. R., "SDWA Amendments: Small Community Compliance", Journal of the American Water Works Association, Vol. 80, No. 8., Pages 38-39 (1988).

Midwest Plan Service, *Private Water Systems Handbook*, Iowa State University, Ames, IA (1979).

Miller, G. W., Cromwell III, J. E., and Marrocco, F. A., "The Role of the States in Solving the Small System Dilemna", Journal of the American Water Works Association, Vol. 80, No. 8, Pages 32-37 (1988).

National Association of Plumbing, Heating, and Cooling Contractors, *National Standard Plumbing Code*, National Association of Plumbing, Heating, and Cooling Contractors, Washington, D.C. (1971).

National Fire Protection Association, "Water Supply Systems for Rural Fire Protection," *National Fire Codes*, Vol. 8, Boston, MA (1969).

National Sanitation Foundation Testing Laboratory, *Seal of Approval Listing of Plastic Materials, Pipe, Fittings and Appurtenances for Potable Water and Waste Water*, National Sanitation Foundation, Ann Arbor, MI (revised annually).

National Water Well Association, "The Authoritative Primer: Ground Water Pollution," *Water Well Journal*, Special Issue, Vol. 24, No. 7 (1970).

New York State Department of Health, *Training Manual for Water Distribution Operations Grade D*, Bureau of Public Water Supply Protection (undated).

Okun, D. A., and Ernst, W. R., *Community Piped Water Supply Systems in Developing Counties*. The World Bank, Washington, D.C. (1987).

Olin Corporation, "Hypochlorination of Water," Olin Corporation - Chemicals Division, New York, NY (1962).

Renner, R. C., "Survival Tactics for Small Systems", Waterworld News, Vol. 6, No. 2, Pages 10-12 (1990).

Sagraves, B. R., Peterson, J. H., and Williams, P. C., "Financing Strategies for Small Systems", Journal of the American Water Works Association, Vol. 80, No. 8, Pages 40-43 (1988).

Salvato, J.A., *The Design of Small Water Systems*, New York State Department of Health, Health Education Service, Albany, NY (1983).

Scharfenaker, Mark A., "AWWA Small Systems Program", Journal of the American Water Works Association, Vol. 80, No. 8, Pages 44-47 (1988).

Smalley, G. , "Networking: A Weapon for Small Systems", Waterworld News, Vol. 6, No. 2., Pages 14-15 (1990).

Sykes, R. G., and Doty, R. N., *"The New Regulations - A Challenge for Small Systems"*, Journal of the American Water Works Association, Vol. 80, No. 8, Pages 62-64 (1988).

Tardiff, R.D., and McCabe, L.J., "Rural Water Quality Problems and the Need for Improvement," *Second Water Quality Seminar Proceedings*, Pages 344-36, American Society of Agricultural Engineers, St. Joseph, MI (1968).

Toft, P., Tobin, R. S., and Sharp. J., "*Drinking Water Treatment Small System Alternatives*, Pergamon Press, Elmsford, NY (1989).

U.O.P. Johnson Division, *Ground Water and Wells*, U.O.P. Johnson Division, St. Paul, MN (1972).

U.S. Agency for International Development, WASH Technical Report No. 61, "Pump Selection: A Field Guide for Developing Counties", Water and Sanitation for Health Project, Washington, D.C. (1989).

U.S. Congress, Office of Technology Assessment, "Using Desalination Technologies for Water Treatment", U.S. Government Printing Office, Washington, D.C. (1988).

U.S. Department of Health and Human Services, Centers for Disease Control, "Water Fluoridation", Atlanta, GA (1986).

U.S. Department of Health and Human Services, Indian Health Service, "Nationwide Indian Health Service Federal Photovoltaic Utilization Program", Washington, D.C. (1990).

U.S. Department of the Interior, Geological Survey, "A Primer on Ground Water," U.S. Government Printing Office, Washington, D.C. (Reprinted annually).

U.S. Environmental Protection Agency, "Fluoridation Engineering Manual," Environmental Protection Agency, Water Supply Division, Washington, D.C. (1972).

U.S. Environmental Protection Agency, *Manual of Water Well Construction Practices*, Environmental Protection Agency, Water Supply Division, Washington, D.C. (1977).

U.S. Environmental Protection Agency, *Cross-Connection Control Manual*, Environmental Protection Agency, Office of Water, Washington, D.C. (1989).

U.S. Environmental Protection Agency, "Guidance Manual for Compliance with The Filtration and Disinfection Requirements for Public Water Systems Using Surface Water Sources", Environmental Protection Agency, Office of Drinking Water, Washington, D.C. (1989).

U.S. Environmental Protection Agency, "Fact Sheet on Home Drinking Water Treatment", Environmental Protection Agency, Office of Water, Washington, D.C. (1989).

U.S. Environmental Protection Agency, "Self Assessment for Small Publicly Owned Water Systems", Environmental Protection Agency, Office of Water, Washington, D.C. (1989).

U.S. Environmental Protection Agency, "A Water and Wastewater Manager's Guide for Staying Financially Healthy", Environmental Protection Agency, Office of Water, Washington, D.C. (1989).

U.S. Environmental Protection Agency, "Environmental Pollution Control Alternatives: Drinking Water Treatment for Small Water Treatment Facilities", Environmental Protection Agency, CERI, Cincinnati, OH (1990).

U. S. Environmental Protection Agency, *Technologies for Upgrading Existing or Designing New Drinking Water Treatment Facilities*, Environmental Protection Agency, Office of Drinking Water, Cincinnati, OH (1990).

U.S. Environmental Protection Agency, "Fact Sheet, Drinking Water Regulations Under the Safe Drinking Water Act", Environmental Protection Agency, Criteria and Standards Division, Office of Drinking Water, Washington, D.C. (1990).

Water Systems Council, *Water Systems Handbook*, 6th edition, Water Systems Council, Chicago, IL (1977).

Whitsell, W.J., and Hutchinson, G.D., "Seven Danger Signals for Individual Water Supply Systems," *Transactions of the American Society of Agricultural Engineers*, American Society of Agricultural Engineers, St. Joseph, MI (1990).

World Health Organization, *Guidelines for Drinking Water Quality: Vol. 3 -Drinking Water Quality Control in Small Community Supplies*, Geneva, Switzerland (1985).

World Health Organization, Technical Paper Series (TP 23), *Renewable Energy Sources for Rural Water Supply*, WHO Collaborating Centre, The Hague, The Netherlands (1986).

World Health Organization, Technical Paper Series, *Slow Sand Filtration for Community Water Supply*, WHO Collaborating Center, The Hague, The Netherlands (1987).

World Health Organization, Technical Paper Series (TP 18), *Small Community Water Suppliers; Technology of Small Water Supply Systems in Developing Countries*, WHO Collaborating Center, The Hague, The Netherlands, (1987).

World Health Organization, *Disinfection of Rural and Small Community Water Supplies*, Copenhagen, Denmark (1989).

Wright, F. B., *Rural Water Supply and Sanitation,* 3rd edition, Robert E. Krieger Publishing Company, Huntington, NY (1977).

APPENDIX A

Health Effects, Source of Contaminants, and Treatment

Contaminant	Health Effects	Contaminant Sources
Arsenic	Carcinogenic	Geological, pesticides, pressure treated wood
Asbestos	Benign tumors	Geological, asbestos-cement pipe
Barium	Circulatory system effects	Geological, oil and gas exploration and production
Cadmium	Kidney effects	Geological, mining, smelting, and corrosion of galvanized pipe
Chromium	Gastrointestinal effects	Geological, faucets
Fluoride	Crippling Fluorosis	Geological, mining
Lead	Gastrointestinal distress, fatigue, anemia, paralysis, and brain damage	Geological, lead service lines, solder and brass faucets
Mercury	Kidney effects	Used in manufacture of paint, paper, vinyl chloride; used in fungicides; geological
Nitrate	Methemoglobinemia ("blue baby" syndrome)	Fertilizer, sewage, feedlots
Nitrite	Methemoglobinemia ("blue baby" syndrome)	Fertilizer, sewage, feedlots
Radioactivity	Bone cancer	Geological, mining, hospital and nuclear waste

Contaminant	Health Effects	Contaminant Sources
Selenium	Neurological effects	Geological, mining, pesticides
Total Trihalomethanes	Carcinogenic	Chlorination of drinking water, bleached paper wastewaters organic chemical industry
Turbidity	Adverse health effects from harbored or shielded bacteria	Particulate from soil, organisms and biomass
	Organics (Solvents)	
Benzene	Carcinogenic	Solvent
cis-1,2-Dichloroethylene	Nervous system, liver, kidney	Extraction solvent, dyes, perfumes, pharmaceuticals, lacquers
1,2-Dichloropropane	Liver toxin, lung, and kidney effects	Pesticide, solvent
Ethylbenzene	Liver, kidney effects	Manufacture of styrene
Monochlorobenzene	Respiratory, nervous system, liver, kidney effects	Solvent, pesticide
o-Dichlorobenzene	Nervous system, lung, liver, kidney effects	Industrial solvent, pesticide
Styrene	Possible cancer, liver, central nervous system effects	Manufacture of polystyrene, plastic
Tetrachloroethylene	Probable cancer	Dry-cleaning solvent
Toluene	Nervous system, lung, liver effects	Solvent

Contaminant	Health Effects	Contaminant Sources
trans-1,2 Dichloroethylene	Nervous system, liver, kidney effects	Extraction solvent, dyes, perfumes, pharmaceuticals, lacquers
Xylenes (total)	Central nervous system effects	Solvent; used to manufacture paint, dyes, adhesives, detergents
	Pesticides/Herbicides/PCBs	
Alachlor	Probable cancer	Herbicide
Aldicarb	Nervous system toxicity	Pesticide, herbicide, restricted in some areas
Aldicarb sulfone	Nervous system toxicity	Pesticide, herbicide, restricted in some areas
Aldicarb sulfoxide	Nervous system toxicity	Pesticide, herbicide, restricted in some areas
Atrazine	Nervous system, liver, heart effects	Herbicide
Carbofuran	Nervous system, reproductive effects	Pesticide, herbicide
Chlordane	Nervous system, liver effects	Pesticide, herbicide, most uses banned in 1980
Dibronochloro-propane (DBCP)	Probable cancer	Pesticide, canceled in 1977
2,4-D	Liver, kidney effects	Herbicide
Endrin	Probable cancer	Insecticide

Contaminant	Health Effects	Contaminant Sources
Ethylenedibromide (EDB)	Probable cancer	Gasoline additive, soil fumigant, solvent, most pesticide uses restricted in 1984
Heptachlor	Probable cancer	Insecticide, most uses restricted in 1983
Heptachlor epoxide	Probable cancer	Insecticide, most uses restricted in 1983
Lindane	Neurological, liver, kidney effects	Insecticide to control fleas, lice, ticks, some uses restricted in 1983
Methoxychlor	Central nervous system effects	Insecticide
PCBs	Probable cancer, reproductive effects	Transformers, capacitors; production banned in 1977
Pentachlorophenol	Organ, central nervous system, fetal effects	Wood preservative; nonwood uses banned in 1987
Toxaphene	Probable cancer	Pesticide, herbicide, most uses canceled in 1977
2,4,5-TP (Silvex)	Liver, kidney effects	Herbicide, canceled in 1983

Best Treatment Methods

Organic Chemicals	**Granular Activated Carbon**	**Packed Tower Aeration**	**Polymer Addition**
Acrylamide			x
Alachlor	x		
Aldicarb	x		
Aldicarb sulfone	x		
Aldicarb sulfoxide	x		
Atrazine	x		
Carbofuran	x		
Calordane	x		
2,4-D	x		
Dibromochloropropane (DBCP)	x	x	
o-Dichlorobenzene	x	x	
cis-1,2-Dichloroethylene	x	x	
trans-1,2-Dichloroethylenc	x	x	
Epichlorohydrin			x
Ethylene dibromide (EDB)	x	x	
Ethylbenzene	x	x	
Heptachlor	x		
Heptachlor epoxide	x		
Lindane	x		
Methoxychlor	x		
Monochlorobenzene	x	x	
PCBs	x		
Pentachlorophenol	x		
Styrene	x	x	
2,4,5-TP (Silvex)	x		
Tetrachloroethylene	x	x	
Toluene	x	x	
Toxaphene	x		
Xylenes (Total)	x	x	

Inorganic Chemicals	**Best Treatment Methods**
Arsenic	Activated Aluminum Coagulation/Filtration Lime Softening Reverse Osmosis
Asbestos	Coagulation/Filtration Corrosion Control Direct and Diatomite Filtration
Barium	Ion Exchange Lime Softening Reverse Osmosis
Cadmium	Coagulation/Filtration Ion Exchange Lime Softening Reverse Osmosis
Chromium	Coagulation/Filtration Ion Exchange (Cation or Anion Depending on Form) Lime Softening (Chromium III Only) Reverse Osmosis
Fluoride	Activated Alumina Reverse Osmosis
Mercury	Coagulation/Filtration Granular Activated Carbon Lime Softening Reverse Osmosis
Nitrate/Nitrite	Ion Exchange Reverse Osmosis
Selenium	Activated Alumina Coagulation/Filtration (Selenium IV Only) Lime Softening Reverse Osmosis

Radioactive Chemicals	**Treatment Method**
Radium	Ion Exchange Lime Softening Oxidation/Filtration Reverse Osmosis
Radon	Aeration Granular Activated Carbon
Uranium	Coagulation/Filtration Ion Exchange (Anion or Cation Depending on pH) Lime Softening Reverse Osmosis
Gross Beta	Ion Exchange Reverse Osmosis

Microorganisms	**Treatment Method**
Giardia Lamblia	Disinfection Filtration
Bacteria	Disinfection Filtration
Viruses	Disinfection Filtration

APPENDIX B

Collection and Analysis of Bacteriological Samples

SAMPLING

Water samples must be collected carefully to prevent contamination. The collector should follow these steps:

1. Use a sterile sample bottle or other container provided by the laboratory that will examine the sample.
2. Insure that water taps used for sampling are free of aerators, strainers, hose attachments, mixing type faucets, and purification devices.
3. Inspect the outside of the faucet. If water leaks around the outside of the faucet, select a different sampling site.
4. Allow the water to run for enough time (about two minutes) to clear the service line before sampling.
5. When filling the bottle, hold bottle so that no water which contacts the hands runs into the bottle. Cap the bottle tightly.
6. Deliver the sample immediately to the laboratory. If possible, store the sample in an iced cooler during transport. In no case should the time between collection and analysis exceed 30 hours. It may be necessary to send a water sample by overnight mail service.

Under EPA's total coliform rule, all water samples must be collected from the distribution system (e.g., household tap), rather than from the well or other location.

ANALYSIS FOR TOTAL COLIFORMS

As stated in Part I of this manual, total coliforms are used to show whether a water supply is contaminated with, or vulnerable to, fecal pollution. EPA is approving four analytical methods for testing water in distribution systems to determine whether total coliforms are present. These are the Membrane Filter Technique, Multiple-Tube Fermentation Technique, Presence-Absence Coliform Test, and the Mixed Media ONPG-MUG test. In all cases, 100-ml water samples are tested; therefore the sample collector must provide at least 100 ml to the laboratory. All microbiology samples which are collected to satisfy the drinking water regulations must be analyzed in a laboratory which is certified by EPA or the State.

In the Membrane Filter Technique, a vacuum pulls 100 ml of water sample through a membrane filter held in place by a filter-holding device. Total coliform and other bacteria are retained on the filter. The filter is then placed on a special medium which allows the growth of total coliform, and incubated at 35°C for 22 to 24 hours. If a total coliform-like colony(ies) is observed on the membrane, the laboratory should make sure that it is a total coliform by using another EPA-approved test.

The Multiple Tube Fermentation Technique involves adding the water sample to either a bottle or a set of tubes, each of which contains either lactose broth or lauryl typtose broth and an inverted tube. The bottle or tubes are then incubated at 35°C for 24 to 48 hours. If gas production is observed in an inverted tube after incubation, the sample contains total coliforms. This is confirmed by adding a small volume of culture from the positive tube(s) or bottle into a tube containing brilliant green lactose bile (BGLB) broth and a smaller inverted tube. After incubating for the tube containing BGLB broth for 48 hours, the inverted tube is observed for gas production. Gas production confirms the presence of total coliforms.

In the Presence-Absence Coliform Test, 100 ml of the water sample is added to a bottle containing P-A Broth, and incubated at 35°C for 24 to 48 hours. If the broth becomes yellow-colored, total coliforms are present. This is confirmed in BGLB broth.

The Mixed Media ONPG-MUG (MMO-MUG) test is the simplest of the EPA-approved tests for total coliforms. A 100 ml water sample is added to a bottle or flask containing the MMO-MUG powder, mixed, and incubated at 35°C for 24 hours. The formation of a yellow color denotes the presence of total coliforms.

If a laboratory finds that a high level of non-coliform bacteria are obscuring the total coliform test, it must not use the sample (unless total coliforms are present), and request that the system collect another sample within 24 hours from the same location as the original sample, and have it analyzed for total coliforms.

Under the total coliform rule, the laboratory must test all total coliform-positive samples for the presence of either fecal coliform or *E. coli*. The analytical methods for these bacteria are described elsewhere.

SURFACE WATER ANALYSIS

Under EPA's surface water treatment requirements, a public system using unfiltered surface water must monitor its source water quality. Unlike the total coliform rule, which requires the system to determine only whether total coliform is present or absent in a sample, this rule requires systems to determine the level of total coliforms or fecal coliforms in a sample. For counting of total coliforms, EPA has approved the 5-tube or 10-tube Multiple Tube Fermentation Technique, the Membrane Filter Technique, and the 5-tube MMO-MUG test. For counting fecal coliforms, EPA has approved the Fecal Coliform Membrane Filter Procedure and the Fecal Coliform Test (EC Medium or A-1 Medium).

The surface water treatment rule also allows a system to determine the amount of heterotrophic bacteria at a site within the distribution system, if the system does not find a disinfectant residual at that site. If the concentration does not exceed 500 bacterial colonies per ml, then the system is considered to have a disinfectant residual for legal purposes. The level of heterotrophic bacteria is determined by the Pour Plate Method, which involves transferring a small volume of water sample to a sterile petri dish, and mixing it with a warm nutrient which solidifies below 45°C. This medium is incubated for 48 hours at 35°C and the number of bacterial colonies counted.

Most of the above-cited methods are described in the 17th Edition of "Standard Methods for the Examination of Water and Wastewater" (1988).

APPENDIX C

Identification By Human Senses[1]

A. SENSE OF FEELING

IMPURITY OR CONTAMINANT	SYMPTOM	CAUSE	MEANS OF TREATMENT
Hard Water	Soap curd, and scum in wash basins & bathtub. Whitish scale deposits in pipes, water heater & tea kettle.	Calcium (limestone) and magnesium salts.	Cation exchange water softener.
Grittiness	Abrasive texture to water when washing or residual left in sink.	Excessively fine sand, silt in water.	Sand trap in ultra-filtration.

B. SENSE OF SMELL

IMPURITY OR CONTAMINANT	SYMPTOM	CAUSE	MEANS OF TREATMENT
Odor	Musty, earthy or wood smell.	Generally, harmless organic matter.	Activated carbon filter.
	Chlorine smell.	Excessive chlorination.	Dechlorinate with activated carbon filter.

---

1 This information has been taken principally from a paper titled "Sensitivity: A Key Water Conditioning Skill" by Wes McGowan. The paper was published in Water Technology, September/October 1982.

IMPURITY OR CONTAMINANT	SYMPTOM	CAUSE	MEANS OF TREATMENT
	Rotten egg odor - tarnished silverware.	1. Dissolved hydrogen sulfide gas.	Manganese greensand filter - constant chlorination followed by filtration/ dechlorination.
		2. Presence of sulfate reducing bacteria in raw water.	Constant chlorination followed by activated carbon filter.
	Hot water, rotten egg odor.	Action of magnesium rod in hot water heater.	Remove magnesium rod from heater.
	Detergent odor, water foams when drawn.	Seepage of septic discharge into underground water supply.	1. Locate and eliminate source of seepage - then heavily chlorinate well.
			2. Activated carbon filter will adsorb limited amount.
	Gasoline or oil (hydro-carbon) smell.	Lack in fuel oil tank or gasoline tank seeping into water supply.	No residential treatment. Locate and eliminate seepage.
	Methane gas.	Naturally occurring caused by decaying organics.	Aeration system and repump.

IMPURITY OR CONTAMINANT	SYMPTOM	CAUSE	MEANS OF TREATMENT
	Phenol smell (chemical odor).	Industrial waste seeping into surface or ground well supplies.	Activated carbon filter will adsorb short-term.

C. SENSE OF TASTE

IMPURITY OR CONTAMINANT	SYMPTOM	CAUSE	MEANS OF TREATMENT
Taste	Salty of brackish	High sodium content	1. Deionize drinking water only with disposable mixed bed - anion/cation resins; or 2. Reverse osmosis; or 3. Home distillation system.
	Alkali taste	High dissolved mineral containing alkalinity. (Stained aluminum cookware.)	1. Reduce by reverse osmosis.
	Metallic taste	1. Very low pH water (3.0-5.5). 2. Heavy iron concentration in water above 3.0 ppm Fe.	1. Correct with calcite type filter (see Acid Water). 2. (See Iron Water).

D. SENSE OF SIGHT

IMPURITY OR CONTAMINANT	SYMPTOM	CAUSE	MEANS OF TREATMENT
Turbidity	Dirt, salt, clay	Suspended matter in surface water pond, stream or lake.	"Calcite" or Neutralite (media) type filter - up to 50 ppm
	Sand grit, silt or clay substances	Well sand from new well or defective well screen.	Sand trap and/or new well screen
	Rust in water	Acid water causing iron "pick-up".	Neutralizing calcite filter to correct low pH acidity and remove precipitated iron
	Grey string-like fiber	Organic mater in raw water algae, etc.	Constant chlorination followed by activated carbon filter to dechlorinate
Acid Water	Green stains on sinks and silver porcelain bathroom fixtures. Blue-green cast to water.	Water which has high carbon dioxide content (pH below 6.8) reacting with brass and copper pipes and fittings.	1. Neutralizing calcite filter down to pH of 5.5, or 2. Calcite/ Magnesia - oxide mix (5 to 1) for higher flow rate and to correct very low pH water. 3. Soda ash chemical feed followed by filtration.

IMPURITY OR CONTAMINANT	SYMPTOM	CAUSE	MEANS OF TREATMENT
Discolored water red, "Iron" water	Brown-red stains on sinks and other porcelain bathroom fixtures. Water turns brown-red in cooking or upon heating. Clothing becomes discolored.	1. Dissolved iron in influent (more than 0.3 ppm Fe^{+}) water appears clear when first drawn at cold water faucet. Above 0.3 ppm Fe causes staining.	1. Can remove 0.5 ppm of Fe^{+} for every grain/gal. of hardness to 10 ppm with water softener and minimum pH of 6.7. 2. Over 10 ppm Fe^{+} chlorination with sufficient retention tank time for full oxidation followed by filtration/ dechlorination. 3. In warm climates residential aerator and filtration will substantially reduce iron content.
		2. Precipitate iron (water not clear when drawn).	1. Up to 10 ppm iron removed by Manganese Greensand filter, if pH 6.7 or higher, or 2. Manganese treated, non-hydrous aluminum silicate filter where pH of 6.8 or higher and oxygen is 15% of total iron content.

IMPURITY OR CONTAMINANT	SYMPTOM	CAUSE	MEANS OF TREATMENT
			3. Downflow water softener with good backwash, up to 1.0 ppm Fe. Above 1 ppm to 10 ppm use calcite filter followed by downflow water softener.
			Calcite media type filter to remove precipitated iron.
	Brownish cast does not precipitate.	Iron pick-up from old pipe with water having a pH below 6.8.1. Organic (bacterial) iron.	1. Treat well to destroy iron bacteria with solution of hydrochloric acid then constant chlorination followed by activated carbon media filtration and dechlorination.
			2. Potassium permanganate chemical feed followed by filtration.
	Reddish color in water sample after standing 24 hours.	Colloidal iron.	Constant chlorination followed by activated carbon media filter dechlorination.

IMPURITY OR CONTAMINANT	SYMPTOM	CAUSE	MEANS OF TREATMENT
Yellow water	Yellowish cast to water after softening and/or filtering.	1. Tannins (humic acids) in water from peaty soil and decaying vegetation.	1. Adsorption via special macro-porous Type I anion exchange resin regenerated with salt (NaC1) up to 3.0 ppm.
			2. Manganese greensand or manganese treated sodium alumino-silicate under proper set of conditions.
Milky water	Cloudiness of water when drawn.	1. Some precipitant sludge created during heating of water.	1. Blow down domestic or commercial hot water heater tank periodically.
		2. High degree of air in water from poorly functioning pump.	2. Water will usually clear quickly upon standing.
		3. Excessive coagulant-feed being carried through filter.	3. Reduce coagulant quantity being fed, service filters properly.
Very high chloride content in water	Blackening and pitting of stainless steel sinks and stainless ware in commercial dishwashers.	1. Excessive salt content.	1. Use other chloride resistant metals.
		2. High temperature drying creates chloride concentration accelerating corrosion.	2. Reduce T.D.S. by reverse osmosis.

E. SENSE OF HEARING

IMPURITY OR CONTAMINANT	SYMPTOM	CAUSE	MEANS OF TREATMENT
Excess Fluorides	Yellowish mottled teeth of children. (No visible color, taste, or odor of water.)	F-above 1.0 ppm in natural water supply.	1. Adsorb excess fluoride and reduce to 0.2 ppm with activated alumina media type filter, or 2. Home distillation system for drinking and cooking water, or 3. Remove F- by complete water deionization via disposable mixed bed for drinking water only.
Nitrates	1. No visible color, taste, or odor of water (above 10.0 ppm as N considered health hazard for infants). 2. Rotten egg or sewage smell - water foams.	1. Heavy use commercial fertilizers with residual NO_3 getting into underground water supply. 2. Boiler blowdown of corrosion inhibitors containing nitrates entering surface or underground water supplies.	1. On water less than 3 ppm, remove with strong base Type II anion exchanger, regenerated with NaC1. Get public health analysis. 2. For drinking and cooking water only; reverse osmosis. Limit of nitrate influent to 25 ppm as N, or 3. Home distillation system for drinking and cooking water.

IMPURITY OR CONTAMINANT	SYMPTOM	CAUSE	MEANS OF TREATMENT
		Human or animal waste pollution containing ammonia seepage in water supplies.	4. Eliminate pollution condition. Sterilize well for 24 hours and have public health analysis.
Radioactive Contaminants	Notices by public health. No color, taste or odor.	Atmospheric fallout contamination of surface water supply sources; or stray isotopes getting in water supply from nuclear wastes. Naturally occurring in deep wells.	1. Can remove most all cationic radioactivity with residential cation exchange water softener. 2. For complete treating assuring removal of both anionic and cationic radioactive contaminants treat with mixed bed deionizer. 3. Reverse osmosis.
		Radon gas given off by decaying radium, dissolved in water.	GAC and aeration are effective in removing radon from water.
Heavy Metals lead, zinc, copper cadmium	No visible color, taste, or odor of water.	1. Industrial waste pollution. 2. Corrosion products from piping caused by low pH waters.	1. Reverse Osmosis for drinking and cooking water, or 2. Complete removal via disposable mixed

IMPURITY OR CONTAMINANT	SYMPTOM	CAUSE	MEANS OF TREATMENT
			bed deionizer for drinking water. 3. Water softener will remove Cu, Cd & Zn under proper conditions - counter current brining suggested.
Arsenic	No visible color, taste, or odor of water. Usually a public health matter.	1. Natural groundwater contaminant in local areas. 2. Industrial waste contaminating water supply. 3. Herbicides containing arsenic.	1. Reverse osmosis will remove up to 90 percent for drinking water. 2. Remove arsenic by complete water deionization using disposable mixed bed; set conductivity meter at 250,000 for exhaustion level for drinking water.
Barium	No visible color, taste, or odor of water. Usually a public health matter (above 1.0 ppm considered health risk).	Naturally occurring in certain geographic regions.	Remove by cation exchange water softener, simultaneously with calcium and magnesium hardness using very strong brine solution.
Boron	Distorted potted plants and chrysanthemums (above 1.0 ppm considered undesirable).	Naturally occurring in southwest and other areas.	Removal via a selective anion exchange resin.

IMPURITY OR CONTAMINANT	SYMPTOM	CAUSE	MEANS OF TREATMENT
Pesticides Herbicides (DDT, 2, 4-D chlordane, etc.)	Sharp chemical taste or odor in water (can be semitoxic).	Excessive agricultural, spraying and water run off to surface supplies.	Activated carbon media filter will adsorb limited amount. Must continue to monitor treated water.
Cyanide	No visible color, taste or odor (above 0.20 ppm considered health risk).	Industrial waste pollution from electroplating, steel & coking.	Continuous chlorination and activated carbon filtration of metals after pH adjustment.
"TCE" contamination (Trichloro-ethylenc)	Notice from Public Health Department	Waste degreasing solution from auto and electric motor clean-up, getting into surface or underground water supply.	Activated carbon filters in series, with constant monitoring between units for break-through.

APPENDIX D

Recommended Procedure for Cement Grouting of Wells for Sanitary Protection[2]

The annular open space on the outside of the well casing is one of the principal avenues through which undesirable water and contamination may gain access to a well. The most satisfactory way of eliminating this hazard is to fill the annular space with neat cement grout. To accomplish this satisfactorily, careful attention should be given to see that:

1. The grout mixture is properly prepared.
2. The grout material is placed in one continuous mass.
3. The grout material is placed upward from the bottom of the space to be grouted.

Neat cement grout should be a mixture of cement and water in the proportion of one bag of cement (94 pounds) and 5 to 6 gallons of clean water. Whenever possible, the water content should be kept near the lower limit given. Hydrated lime to the extent of 10 percent of the volume of cement may be added to make the grout mix more fluid and thereby facilitate placement by the pumping equipment. Mixing of cement or cement and hydrated lime with the water must be thorough. Up to 5 percent by weight of bentonite clay may be added to reduce shrinkage.

GROUTING PROCEDURE

The grout mixture must be placed in one continuous mass; hence, before starting the operation, sufficient materials should be on hand and other facilities available to accomplish its placement without interruption.

Restricted passages will result in clogging and failure to complete the grouting operation. The minimum clearance at any point, including couplings, should not be less than 1½ inches. When grouting through the annular space, the grout pipe should not be less than 1-inch nominal diameter. As the grout moves upward, it picks up much loose material such as results from caving. Accordingly, it is desirable to waste a suitable quantity of the grout which first emerges from the drill hole.

In grouting a well so that the material will move upward, there are two general procedures that may be followed. The grout pipe may be installed within the well casing or in the annular space between the casing and drill hole if there is sufficient clearance to permit this. In the latter case, the grout pipe is installed in the annular space to within a few inches of the bottom. The grout is pumped through this pipe, discharging into the

---

2 This information has been taken principally from a pamphlet of the Wisconsin State Board of Health entitled "Method of Cement Grouting for Sanitary Protection of Wells." The subject is discussed in greater detail in that publication. (NOTE: Publication is out of print.)

surface. In three to seven days the grout will be set, and the well can be completed and pumping started. A waiting period of only 24 to 36 hours is required if quick-setting cement is used.

When the grout pipe is installed within the well casing, the casing should be supported a few inches above the bottom during grouting to permit grout to flow into the annular space. The well casing is fitted at the bottom with an adapter threaded to receive the grout pipe and a check valve to prevent return of grout inside of the casing. After grout appears at the surface, the casing is lowered to the bottom and the grout pipe is unscrewed immediately and raised a few inches. A suitable quantity of water should then be pumped through it, thereby flushing any remaining grout from it and the casing. The grout pipe is then removed from the well and three to seven days are allowed for setting of the grout. The well is then cleared by drilling out the adapter, check valve, plug, and grout remaining within the well.

A modification of this procedure is the use of the well casing itself to convey the grout to the annular space. The casing is suspended in the drill hole and held several feet off the bottom. A spacer is inserted in the casing. The casing is then capped and connection made from it to the grout pump. The estimated quantity of grout, including a suitable allowance for filling of crevices and other voids, is then pumped into the casing. The spacer moves before the grout, in turn forcing the water in the well ahead of it. Arriving at the lower casing terminal, the spacer is forced to the bottom of the drill hole, leaving sufficient clearance to permit flow of grout into the annular space and upward through it.

After the desired amount of grout has been pumped into the casing, the cap is removed and a second spacer is inserted in the casing. The cap is then replaced and a measured volume of water sufficient to fill all but a few feet of the casing is pumped into it. Thus all but a small quantity of the grout is forced from the casing into the annular space. From three to seven days are allowed for setting of the grout. The spacers and grout remaining in the casing and drill hole are then drilled out and the well completed.

If the annular space is to be grouted for only part of the total depth of the well, the grouting can be carried out as directed above when the well reaches the desired depth, and the well can then be drilled deeper by lowering the tools inside of the first casing. In this type of construction, where casings of various sizes telescope within each other, a seal should be placed at the level where the telescoping begins, that is, in the annular space between the two casings. The annular space for grouting between two casings should provide a clearance of at least 1½ inches, and the depth of the seal should be not less than 10 feet.

APPENDIX E

Emergency Disinfection

When ground water is not available and surface water must be used, avoid sources containing floating material or water with a dark color or an odor. The water tank from a surface source should be taken from a point upstream from any inhabited area and dipped, if possible, from below the surface.

When the home water supply system is interrupted by natural or other forms of disaster, limited amounts of water may be obtained by draining the hot water tank or melting ice cubes.

In case of a nuclear attack, surface water should not be used for domestic purposes unless it is first found to be free from excessive radioactive fallout. The usual emergency treatment procedures do not remove such substances. Competent radiological monitoring services as may be available in local areas should be relied upon for this information.

There are two general methods by which small quantities of water can be effectively disinfected. One method is by boiling. It is the most positive method by which water can be made bacterially safe to drink. Another method is chemical treatment. If applied with care, certain chemicals will make most waters free of harmful or pathogenic organisms.

When emergency disinfection is necessary, the physical condition of the water must be considered. The degree of disinfection will be reduced in water that is turbid. Turbid or colored water should be filtered through clean cloths or allowed to settle, and the clean water drawn off before disinfection. Water prepared for disinfection should be stored only in clean, tightly covered, noncorrodible containers.

METHODS OF EMERGENCY DISINFECTION

1. *Boiling.* Vigorous boiling for one minute will kill any disease-causing microorganisms present in water. The flat taste of boiled water can be improved by pouring it back and forth from one container into another, by allowing it to stand for a few hours, or by adding a small pinch of salt for each quart of water boiled.
2. *Chemical Treatment.* When boiling is not practical, chemical disinfection should be used. The two chemicals commonly used are chlorine and iodine.
   a. *Chlorine*
      (1) *Chlorine Bleach.* Common household bleach contains a chlorine compound that will disinfect water. The procedure to be followed is usually written on the label. When the necessary procedure is not given, one should find the percentage of available chlorine on the label and use the information in the following tabulation as a guide:

Avalible chlorine[1]	Drops per quart of clear water[2]
1%	10
4-6%	2
7-10%	1

[1] If strength is unknown, add 10 drops per quart of water.
[2] Double amount for turbid or colored water.

The treated water should be mixed thoroughly and allowed to stand for 30 minutes. The water should have a slight chlorine odor; if not, repeat the dosage and allow the water to stand for an additional 15 minutes. If the treated water has too strong a chlorine taste, it can be made more palatable by allowing the water to stand exposed to the air for a few hours or by pouring it from one clean container to another several times.

(2) *Granular Calcium Hypochlorite.* Add and dissolve one heaping teaspoon of high-test granular calcium hypochlorite (approximately 1/4 ounce) for each 2 gallons of water. This mixture will produce a stock chlorine solution of approximately 500 mg/L, since the calcium hypochlorite has an available chlorine equal to 70 percent of its weight. To disinfect water, add the chlorine solution in the ratio of one part of chlorine solution to each 100 parts of water to be treated. This is roughly equal to adding 1 pint (16 oz.) of stock chlorine solution to each 12.5 gallons of water to be disinfected. To remove any objectionable chlorine odor, aerate the water as described above.

(3) *Chlorine Tablets.* Chlorine tablets containing the necessary dosage for drinking water disinfection can be purchased in a commercially prepared form. These tablets are available from drug and sporting goods stores and should be used as stated in the instructions. When instructions are not available, use one tablet for each quart of water to be purified.

b. *Iodine*

(1) *Tincture of Iodine.* Common household iodine from the medicine chest or first aid package may be used to disinfect water. Add five drops of 2 percent United States Pharmacopeia (U.S.P.) tincture of iodine to each quart of clear water. For turbid water add ten drops and let the solution stand for at least 30 minutes.

(2) *Iodine Tablets.* Commercially prepared iodine tablets containing the necessary dosage for drinking water disinfection can be purchased at drug and sporting goods stores. They should be used as stated in the instructions. When instructions are

not available, use one tablet for each quart of water to be purified.

Water to be used for drinking, cooking, making any prepared drink, or brushing the teeth should be properly disinfected.

APPENDIX F

State Drinking Water Agencies

Name/Address

Water Supply Branch
Department of Environmental Management
1751 Federal Drive
Montgomery, Alabama 36130

Alaska Drinking Water Program
Water Quality Management
Department of Environment Conservation
Pouch O
Juneau, Alaska 99811

Manager, Compliance Unit
Waste and Water Quality Management
Room 202
2005 North Central Avenue
Phoenix, Arizona 85004

Division of Engineering
Arkansas State Department of Health
4815 West Markham Street
Little Rock, Arizona 72201

Sanitary Engineering Branch
California Department of Health
714 P Street
Sacramento, California 95814

Drinking Water Section
Colorado, Department of Health
4210 East 11th Avenue
Denver, Colorado 80220

Water Supplies Section
Connecticut Department of Health
79 Elm Street
Hartford, Connecticut 06115

Name/Address

Program Director
Office of Sanitary Engineering
Division of Public Health
Jesse Cooper Memorial Building
Capital Square
Dover, Delaware 19901

Drinking Water Section
Department of Environmental Regulation
2600 Blair Stone Road
Tallahassee, Florida 32301

Program Manager
Department of Natural Resources
270 Washington Street, SW
Atlanta, Georgia 30334

Drinking Water Program
Sanitation Branch
Environmental Protection and Health Services Division
P.O. Box 3378
Honolulu, Hawaii 96801

Bureau of Water Quality
Division of Environment
Idaho Department of Health and Welfare Statehouse
Boise, Idaho 83720

Division of Public Water Supplies
Illinois Environmental Protection Agency
2200 Churchill Road
Springfield, Illinois 62706

Name and Address

Water Supply Division
Department of Water, Air and
Waste Management
Wallace State Office Building
Des Moines, Iowa 53019

Division of Water
Department of Environmental
Protection
18 Reilly Road, Fort Boone Plaza
Frankfort, Kentucky 40601

Support Services Section
Kansas Division of Environment
Forbes Field
Topeka, Kansas 66620

Office of Preventive and Public
Health Services
Louisiana Department of Health
and Human Services
P.O. Box 60630
New Orleans, Louisiana 70160

Division of Water Supply
Inspection & Compliance Program
Department of Health and Mental
Hygiene
Office of Environmental Programs
201 West Preston Street
Baltimore, Maryland 21201

Division of Water Supply
Department of Environmental
Quality Engineering
One Winter Street
Boston, Massachusetts 02108

Assistant Director
Department of Human Services
Bureau of Health
Division of Health Engineering
State House
Augusta, Maine 04333

Name and Address

Action Director
Division of Public Water Supply
Indiana State Board of Health
1330 West Michigan Street
Indianapolis, Indiana 46202

Water Supply Services Division
Environmental and Occupational
Health Services Administration
3500 North Logan Street
Lansing, Michigan 48909

Section of Public Water Supplies
Minnesota Department of Health
717 Delaware Street
Minneapolis, Minnesota 55440

Division of Water Supply
State Board of Health
P.O. Box 1700
Jackson, Mississippi 39205

Public Drinking Water Program
Division of Environmental Quality
P.O. Box 1368
Jefferson City, Missouri 65102

Bureau of Water Quality
Health and Environmental Services
Cogswell Building, Room A206
Helena, Montana 59620

Division of Environmental Health
and Housing Surveillance
Nebraska Department of Health
301 Sentenial Mall South
Lincoln, Nebraska 68509

Water Supply Division
New Hampshire Water Supply and
Pollution Control Commission
P.O. Box 95, Hazen Drive
Concord, New Hampshire 03301

Name and Address

Bureau of Potable Water
Division of Water Resources
New Jersey Department of
Environmental Protection CN-029
Trenton, New Jersey 08625

Health Program Manager
Water Supply Section
Environmental Improvement Division
P.O. Box 968
Santa Fe, New Mexico 87504-0968

Bureau of Public Water Supply
Protection
State of New York Department of
Health
Office of Public Health
Tower Building
Nelson A. Rockefeller Empire State
Plaza
Albany, New York 12237

Water Supply Branch
Division of Health Services
P.O. Box 2091
Raleigh, North Carolina 27602-2091

Division of Water Supply
and Pollution Control
State Department of Health
1200 Missouri Avenue
Bismarck, North Dakota 58501

Office of Public Water Supply
Ohio Environmental Protection
Agency
361 East Broad Street
P.O. Box 1049C
Columbus, Ohio 43216

Name and Address

Water Facility Engineering Service
Oklahoma State Department of
Health
P.O. Box 53551
Oklahoma City, Oklahoma 73152

Drinking Water Systems
Department of Human Resources
Health Division
1400 SW 5th Avenue
Portland, Oregon 97201

Bureau of Water Supplies
Department of Environmental
Resources
P.O. Box 2063
Harrisburg, Pennsylvania 17120

Drinking Water Supply Supervision
Program
Puerto Rico Department of Health
P.O. Box 70184
San Juan, Puerto Rico 00936

Division of Water Supply
Rhode Island Department of Health
75 Davis Street, Health Building
Providence, Rhode Island 02908

Division of Water Supply
Department of Health and
Environmental Control
2600 Bull Street
Columbia, South Carolina 29201

Bureau of Drinking Water
Water and Natural Resources
Joe Foss Building
523 Capital Avenue, East
Pierre, South Dakota 57501

Name and Address

Bureau of Public Water Supplies
Utah State Department of Health
560 South 300 East
Salt Lake City, Utah 84111

Vermont Department of Health
60 Main Street
P.O. Box 70
Burlington, VT 05402

Division of Water Supply Engineering
Virginia State Department of Health
James Madison Building
109 Governor Street
Richmond, Virginia 23219

Water Supply and Waste Section
Department of Social and Health Services
Mail Stop LD-11
Olympia, Washington 98504

Drinking Water Division
Office of Environmental Health Services
State Department of Health
1800 Washington Street E
Charleston, WV 25305

Bureau of Water Supply
Public Water Supply Section
Department of Natural Resources
P.O. Box 7921
Madison, Wisconsin 53707

Water Quality Division
Department of Environmental Quality
401 West 19th Street
Cheyenne, Wyoming 82002

Name and Address

Guam Environmental Protection Agency
Government of Guam
P.O. Box 2999
Agana, Guam 96910

Public Health Engineering
Nevada Department of Human Resources
Consumer Health Protection Services
505 East King Street, Room 103
Carson City Nevada 89710

Division of Water Supply
Tennessee Department of Health and Environment
150 9th Avenue, North
Nashville, Tennessee 37219-5404

Division of Water Hygiene
Texas Department of Health
1100 West 49th Street
Austin, Texas 78756

Division of Environmental Quality
Commonwealth of the Northern Mariana Islands
P.O. Box 1304
Saipan, CM 96950

Environmental Protection Board HQ
Trust Territories of Pacific Island
Saipan, CM 96950

Department of Conservation and Cultural Affairs
Government of Virgin Islands
P.O. Box 4340
St. Thomas, Virgin Islands 00801

APPENDIX G

EPA Regional Offices

Name/Adresses

USEPA Region I - Connecticut, Maine, Massachusetts, New Hampshire, Rhode Island, Vermont
John F. Kennedy Federal Building
Boston, MA 02203
(617) 223-5731

Region II - New Jersey, New York, Puerto Rico, Virgin Islands
Federal Building
26 Federal Plaza
New York, NY 10007
(212) 264-1800

Region III - Delaware, District of Columbia, Maryland, Pennsylvania, Virginia, West Virginia
Curtis Bldg., 6th & Walnut Streets
Philadelphia, PA 19106
(215) 597-8227

Region IV - Alabama, Florida, Georgia, Kentucky, Mississippi, North Carolina, South Carolina, Tennessee
345 Courtland Street
Atlanta, GA 30308
(404) 881-3781

Region V - Illinois, Indiana, Michigan, Minnesota, Ohio Wisconsin
230 South Dearborn Street
Chicago, IL 60604
(312) 353-2151

Name/Addresses

Region VI - Arkansas, Louisiana, New Mexico, Oklahoma, Texas
1201 Elm Street
Dallas, TX 75270
(214) 749-2106

Region VII - Iowa, Kansas, Missouri, Nebraska
324 E. 11th Street
Kansas City, MO 64106
(816) 374-5429

Region VIII - Colorado, Montana, North Dakota, South Dakota, Utah, Wyoming
Lincoln Tower Building
1860 Lincoln Street
Denver, CO 80295
(303) 837-2731

Region IX - Arizona, California, Hawaii, Nevada, Guam, American Samoa, Trust Territory of Pacific Islands
215 Fremont Street
San Francisco, CA 94105
(415) 974-8106

Region X - Alaska, Idaho, Oregon, Washington
1200 Sixth Avenue
Seattle, WA 98101
(206) 442-1223

APPENDIX H

National Organizations

American Water Resources Association
5410 Grosvenor Lane
Suite 220
Bethesda, Maryland 20814
Phone: (301) 493-8600

American Water Works Association
6666 West Quincy Avenue
Denver, Colorado 80235
Phone: (303) 794-7711

Association of Boards of Certification for Operating Personnel
P.O. Box 786
Ames, Iowa 50010
Phone: (515) 232-3623

Chlorine Institute
2001 L. Street, N.W.
Suite 506
Washington, D.C. 20036
Phone: (202) 775-2790

National Lime Association
3601 N. Fairfax Drive
Arlington, Virginia 22201
Phone: (703) 243-5463

National Rural Water Association
2915 South Thirteenth Street
P.O. Box 1428
Duncan, Oklahoma 73534
Phone: (405) 252-0629

National Sanitation Foundation
P.O. Box 1468
Ann Arbor, Michigan 48106
Phone: (313) 769-8010

National Small Flows Clearinghouse
258 Stewart Street
P.O. Box 6064
Morgantown, WV 26506-6064
Phone: (304) 293-4191

National Water Supply Improvement Association
P.O. Box 102
St. Leonard, Maryland 20685
Phone (301) 855-1173

National Water Well Association
6375 Riverside Drive
Dublin, Ohio 43017
Phone: (614) 761-1711

Rural Community Assistance Program
602 South King Street
Suite 402
Leesburg, Virginia 22075
Phone: (703) 771-8636

Water Systems Council
600 South Federal Street
Suite 400
Chicago, Illinois 60605
Phone: (312) 922-6222

APPENDIX I

Farmers Home Administration

FmHA State Director
Room 717, Aronov Building
474 South Court Street
Montgomery, Alabama 36104
205-832-7077

FmHA State Director
P.O. Box 1289
Palmer, Alaska 99645
907-745-2176

FmHA State Director
Room 3433, Federal Building
230 North First Avenue
Phoenix, Arizona 85025
602-261-6701

FmHA State Director
Room 5529, Federal Office Building
700 West Capitol
P.O. Box 72203
Little Rock, Arkansas 72201
501-378-6281

FmHA State Director
Cleveland Street
Woodland, California 95695
916-666-3382

FmHA State Director
Room 231, 2490 West 26th Avenue
Denver, Colorado 80211
303-837-4347

FmHA State Director
151 East Chestnut Hill Road
Suite 2
Newark, Delaware 19713
302-573-6694

FmHA State Director
401 S.E. 1st Avenue
Room 314 Federal Building
P.O. Box 1088
Gainseville, Florida 32602
904-376-3218

FmHA State Director
Stephens Federal Building
355 E. Hancock Avenue
Athens, Georgia 30601
404-546-2162

FmHA State Director
345 Kekuanaoa Street
Hilo, Hawaii 96720
808-961-4781

FmHA State Director
Room 429
304 N. Eighth Street
Boise, Idaho 83702
208-334-1301

FmHA State Director
2106 W. Springfield Avenue
Champaign, Illinois 61820
217-398-5235

FmHA State Director
5610 Crawfordsville Road
Suite 1700
Indianapolis, Indiana 46224
317-248-4442

Midwest State Director at Large
Room 176, Federal Building
444 SE Quincy Street
Topeka, Kansas 66683
913-295-2870

FmHA State Director
USDA Office Building
Orono, Maine 04473
207-866-4928

FmHA State Director
One Vahlsing Center
Robbinsville, New Jersey 08691
609-259-3136

FmHA State Director
451 West Street
Amherst, Massachusetts 01002
413-253-3471

FmHA State Director
Room 209, Manly Miles Building
1405 South Harrison Road
East Lansing, Michigan 48823
517-337-6631

FmHA State Director
Room 252 Federal Building and
U.S. Courthouse
316 North Robert Street
St. Paul, Minnesota 55101
612-725-5842

FmHA State Director
Suite 831, Federal Building
Jackson, Mississippi 39201
601-960-4316

FmHA State Director
555 Vandriver Drive
Columbia, Missouri 65201
314-442-2271, Ext. 3241

FmHA State Director
Room 234, Federal Building
P.O. Box 850
Bozeman, Montana 59715
406-587-5271, Ext. 4221

FmHA State Director
Room 308 Federal Building
100 Centennial Mall North
Lincoln, Nebraska 68508
402-471-5551

FmHA State Director
Room 3414, Federal Building
517 Gold Avenue, SW
Albuquerque, New Mexico 87102
505-766-2462

FmHA State Director
Room 871, U.S. Courthouse
and Federal Building
100 South Clinton Street
Syracuse, New York 13202
315-423-5290

FmHA State Director
Room 525
310 New Bern Avenue
Raleigh, North Carolina 27601
919-755-4640

FmHA State Director
Room 208, Federal Building
Third and Rosser
P.O. Box 1737
Bismarck, North Dakota 58502
255-4011, Ext. 4781

FmHA State Director
Room 507, Federal Building
200 North High Street
Columbus, Ohio 43215
614-469-5606

FmHA State Director
USDA Agricultural Center Building
Stillwater, Oklahoma 74074
405-624-4250

FmHA State Director
Room 1590, Federal Building
1220 SW 3rd Avenue
Portland, Oregon 97204
503-221-2731

FmHA State Director
Room 728, Federal Building
P.O. Box 905
Harrisburg, Pennsylvania 17108
717-782-4476

FmHA State Director
Room 623
Federico Degetau Federal Building
Carlos Chardon Street
Hato Rey, Puerto Rico 00818
809-753-4481

FmHA State Director
Storm Thurmond Federal Building
Room 1007
1835 Assembly Street
Columbia, South Carolina 29201
803-765-5876

FmHA State Director
Room 308, Huron Federal Building
200 4th Street, SW
Huron, South Dakota 57350
606-352-8651, Ext. 355

FmHA State Director
538 Federal Building & Courthouse
801 Broadway
Nashville, Tennessee 37203
615-251-7341

FmHA State Director
Room 1005, Federal Building
Capser, Wyoming 82602
307-265-5550, Ext. 5271

FmHA State Director
Suite 102, Federal Building
101 South Main
Temple, Texas 76501
817-774-1301

FmHA State Director
Room 5434, Federal Building
125 South State Street
Salt Lake City, Utah 84138
801-524-5027

FmHA State Director
141 Main Street
P.O. Box 588
Montpelier, Vermont 05602
802-223-2371

FmHA State Director
Room 8213, Federal Building
400 North Eighth Street
P.O. Box 10106
Richmond, Virginia 23240
804-771-2451

FmHA State Director
301 Yakima Street
Wenatchee, Washington 98801

FmHA State Director
75 High Street
Morgantown, West Virginia 26505
304-599-4791

FmHA State Director
1257 Mains Street
Stevens Point, Wisconsin 54481
715-341-5900

INDEX

A

Acidity 15, 80, 85
Activated aluminum . . . .
Activated carbon filters 101, 102
Aeration 91
Air line 56
Air rotary drilling 52
Algae 97-98
Alkalinity 12, 13
Alum 80
Aluminum 13
American Water Works Association
(AWWA) 2, 151
Analysis of water:
bacteriological 22
chemical 22
radiological 22
Aquifer 37
Arsenic 167
Artesian aquifers 9
Average daily water use 22
Artesian wells 38

B

Backwash 83
Bacteria:
in water 37
Barium in water 18, 167
Biological characteristics 15, 16
"Blue baby" disease 14
Bored wells 44, 46

C

Cable tool drilling 50
Cadmium 18
Calcium hypochlorite 85
Calcium 13
Capillary fringe 9
Carbon dioxide 96, 16
Cartridge filters 83
Casings 46,52
Catchments 66-73
Centrifugal pumps 105, 106
Cesspools 34, 37, 39
Check valves 129
Chemical characteristics:
of ground water 12-15
of water 12-15
Chemical disinfection 84
Chlorides 13
Chlorination equipment 87-89
Chlorine Disinfection 84-87
Chromium 18
Churn drill 50
Circuit riders 151
Cisterns 66, 68
Coagulation 80
Coliforms 20
Color 11
Community Development Block Grants
(CDBG) 154
Community Involvement 2
Community Planning and Development
Agency 154
Computers 157
Cone of depression 39
Conservation 24, 25
Consolidated formations 34
Contact time 85
Contamination:
physical 21
chemical 22
biological 22
radiological 22
distances to sources 35-36
evaluating treats 36-38
sources of 21
Contract services 160
Copper 13
Copper sulfate 97
Corrosion 96

Corrosion control 95-96
County utilities 160
Crib 44
Cross-connection 136-138
Cryptosoridium 16

D

DPD colorimetric test 87
Diatomaceous earth filters 83
Disinfection: 19
emergency 191
wells 59-62
with ozone 90
Distribution system 28-29
Down-the-hole air hammer 52
Development of groundwater . . 38-42
Development of springs 74-76
Drainage area 69
Drawdown 39
Drilled wells 62
Drilling equipment and methods 44-57
Drive (well) points 48
Driven wells 46-48, 62
Drying beds 103
Dug wells 44, 60

E

Economic Development Administration
(EDA) 154
Electrodialysis reversal 90, 91
Embankment 72
Epsom salts in water 35
Escherichia coli (E. coli) 16, 20
Evaporation rates 7, 8
Evaporation ponds 103, 104
Evapotranspiration 7, 8

F

Farmers Home Administration
(FmHA) 2, 153, 159
Filters 80-83
Filtration, surface water 19
Fluoride removal 94-95
Financial assistance 2
Fire protection 25-27
Flocculation 80
Flooding of wells 58
Flowing artesian wells 62
Fluoride 13
Foaming 11
Formations:
consolidated 34
unconsolidated 34
Freezing protection: 71, 73

G

General litigation bonds 155
Giardia lamblia 16, 65, 83
Glauber's salt 35
Granular Activated Carbon (GAC) 101
Ground level reservoirs 66-69
Ground water:
basins 33-34
temperature 35

H

Hand pump 125-126
Hardness 13
Head loss 139-142
Herbicides 11
Home water needs 24
Household water treatment . . 100-102
Hydraulic ram 105, 111
Hydraulic rotary drill 50
Hydrogen sulfide 95

I

Industrial development bonds . . . 155
Infiltration galleries 64, 76-77
Information Resources 149-150
Inspection:
of pitless installations 134
of wells 55
Interference 41
Iodine 192
Ion exchange 93-94

Iron bacteria 92-93
Iron 14, 91

J

Jet pumps 106
Jetted wells 48-50

L

Lakes 69-71, 73
Lawn sprinkling 25
Lead 14, 37
Leak detection 24
Lead levels 19
Lead control 21
Lightning, protection 126
Lime 96
Lime-soda ash 94
Limestone 33

M

Man-made radiation 17
Magnesium in water 13
Manganese 14, 92
Maximum Contaminant
Levels (MCL) 17, 19
Membrane filter (MF) 175,176
Monitoring 19
Mud rotary drilling 50, 52

N

Natural radiation 17
National Rural Water Association
(NRWA) 151
National Sanitation Foundation
(NSF) 2,151
National Water Supply Improvement
Association (NWSIA) 2
National Water Well Association . . 34
Nitrates 14
Non-artesian wells 38
NTU 12

O

Odors 11
Operating Ratio 156
Organic material 151, 153
Ozone 90
Organic compounds 37

P

Package plants 98-100
Painting 145
Percussion (cable tool) drilling . . . 50
pH 15
Pesticides 14-15
Photo-Voltaic (PV) array . . . 106-109
Pipe:
fittings, friction loss in 141
for distribution systems 138
for well casing 52
Pitless adapters 128-136
Point of entry treatment (POE) 100-101
Point of use treatment (POU) 100-101
Pollution, sources of 21-22
Ponds and lakes 69-73
Positive displacement pumps 105
Potassium permanganate 100
Pressure 26,139-142
Pressure sand filter 83
Pressure tanks 143-144
Preventive Maintenance . . . 96-97, 137
Protection of distribution
system 145, 148
Pump:
alignment in wells . 113, 117, 121
lubrication 106, 117
platforms . . . .
priming 105, 109, 116, 120
Pumphouses 126, 127
Pumps:
centrifugal 105-106
hand 111, 125-126
helical or spiral rotor 105
installation of 121, 123
jet 106
line-shaft (vertical) turbine 121-123

Regenerative turbine pump . . 105
positive displacement 105
solar photovoltaic
(PV) pump 106, 109, 123
wind pump 109, 100, 124, 125
submersible 106
vertical turbine 106
reciprocating 105

Q

Quantity of water 22

R

Radiation 17, 167
Radioactivity 167
Radiological characteristics 16
Radionuclides 37
Radius of influence 39
Radon 173
Rapid sand filter 82-83
Rates of flow 139
Reciprocating pumps 105
Reconstruction of dug wells . . . 63-64
Reservoir 144
Revenue bonds 155
Reverse osmosis 90
Revolving funds 154
Rock formations 33
Rotary drilling methods 52
Rural Community Assistance Program
(RCAP) 151

S

Safe Drinking Water Act
(SDWA) 1-3, 17-21
Salt 15
Sample collection 21
Sand 48
Sanitary covers:
for spring boxes, cisterns 74
for wells 58
Sanitary protection:
of distribution systems 142
of springs 74
of wells 63
of pumping facilities 124
Sanitary quality of groundwater . . . 34
Sanitary survey 27-31
Satellite operation 158-159
Scale 96
Screens, well:
selection of 53, 54
Seal, cement grout formation 189-190
Secondary Maximum Contaminant
Levels (SMCLs) 17, 19
Sedimentation 80
Selenium 18
Settling basins 80
Silver 15
Skeletal fluorosis 94
Slow sand filters 80-82
Sludge 103
Snow 10
Sodium 15
Softening 93-94
Soil moisture 9
Solar irradiation 123
Solar pumps 106, 109
Surface water 65-77
Specific capacity 41
Springs 39, 64
Staining of clothing and fixtures . . 14
State utilities 160
Static pressure 113
Steel pipe 139
Storage tank 144-145
Storage tanks, painting of . . . 144-145
Stream 10
Submersible pump 106
Sulfates 15
Superchlorination 89
Surface Water Treatment Rule
(SWTR) 20, 65

T

Taste and odor in water . . . 11, 73, 95
Temperature:
of ground water 35

Testing:
pitless adapters and units for leaks
water for bacteria 21
water for minerals 21
Total coliform 16
Total Coliform Rule 20
Total dissolved solids (TDS) 15
Toxic substances 14-15
Transpiration rates 8
Trihalomethanes (THM) 11
Turbidity 10, 12, 83

U

Ultraviolet light 89
Unconsolidated formations 34
U.S. Department of Housing and
Urban Development (HUD) ... 2
U.S. Geological Survey (USGS) .. 55
U.S. Small Business
Administration 2, 154

V

Valves:
check 112, 114, 129
Vent, well 120
Vertical turbine pumps 106
Viruses 65
Volatile Organic Compounds
(VOCs) 91

W

Waste treatment 79-104
Water conditioning 93-94
Water conservation 24-25
conservation methods 24-25
structural methods 24-25
economic methods 25
usage restrictions 25
legal means 25
public awareness 25
Water consumption (demand) 24
Water disinfection:
with chemicals 80, 84-89
with ultraviolet light 89-90
Water districts 159
Water quality 34-35
Water rates 155-157
Water rights 7
Water scaling 96
Water softeners 93-94
Water table 39
Water testing 21
Water treatment 79
Water use 220
Watershed 10
Weed control 98
Well:
abandonment (destruction) ... 63
casing 52
construction 44-63
covers 58-59
development (see also well
construction) 54-55, 58, 64
site selection for wells 35-41
disinfection 59-63
failure 55-57
grouting 59
pits 44
points 49
repair 55
screens 53
seals 58-59
slabs 59
Well testing:
yield a drawdown 55
for capacity 41
Windmill 109
Wellhead protection 20

Y

Yield:
of wells 41-42

Z

Zeolite softening 93-94
Zinc in water 15
Zone of saturation 9

Milton Keynes UK
Ingram Content Group UK Ltd.
UKHW040059071024
449327UK00019B/671